The Tale of the Three Little Pigments

Howard Thomas

Book title: The Tale of the Three Little Pigments
Author: Howard Thomas, www.sidthomas.net/wp

Published by Howard Thomas, www.plantsenescence.org
Printer: Cambrian Printers, Aberystwyth
Publication date: 2018 (first edition)

ISBN 978-0-9954751-3-7

For my mother Joan Thomas, my sister Gillian and my brother Brian

CONTENTS

Once upon a time there were Three Little Pigments. ***This*** *little pigment is a chlorophyll, which makes plants green and is a member of a family of relatives found in every living thing.* ***This*** *little pigment is a carotenoid, golden in colour, whose diverse kith and kin are even more numerous than those of the chlorophylls. And* ***this*** *little pigment is a purple phenylpropanoid, scion of the most varied chemical clan of all. The Three Little pigments and their extended families account for most of the familiar colours of the living world. Their story will begin with a prelude on the theme of light and colour and is followed by a three-act drama which will also include walk-on parts for other minor characters. It ends with an epilogue on the dying of the light. To understand pigments is to gain the most intimate insight into what it means to be a plant, and to be a human in a green world.*

1. SUNLIGHT

Concerning light and colour[1]

As thikke as motes in the sonne-beem

Geoffrey Chaucer (*The Canterbury Tales. Tale of the Wyf of Bathe*)

2

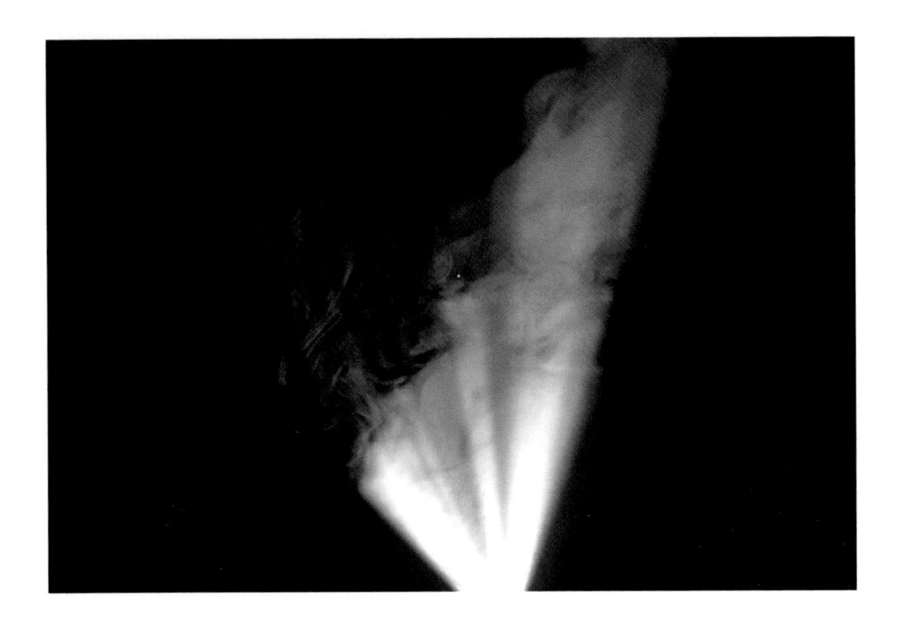

Light is everywhere and nowhere[2]

3

These days smoking in cinemas isn't allowed. You sit there in the dark and watch the moving images before you, projected from the source at the back of the theatre. The space between the projector and the screen must be streaming with photons. That this is so was evident in bygone times, when the air was thick with cigarette smoke, forming a swirling haze in the projector's beam. But nowadays there is no smoke, and the intervening space is darkness. This is a glimpse into the strange world of the quantum and the wave-particle duality of the photon. Photons are manifestly making their way from light source to cinema screen but where, and what, are they between the start of their journey and the destination? The greatest brains in physics have been tormented by this question, and the answer remains unresolved. Near the end of his life, Albert Einstein himself wrote: 'All the fifty years of conscious brooding have brought me no closer to the answer to the question, "What are light quanta?" Of course today every rascal thinks he knows the answer, but he is deluding himself'[3]. As someone once said, a photon isn't a thing but a state: where does your lap go when you stand up?

As it is for the projector and the screen, so it is for the sun and its satellites[4]

5

In the 14 billion degree furnace at the core of our neighbourhood star, the hydrogen-helium fusion cycle converts matter into energy in accordance with the relation $e = mc^2$ (Einstein again). Solar energy radiates out in all directions, filling space with photons across almost the entire electromagnetic spectrum, from short wavelength X-rays through ultraviolet and visible light, to infrared radiation and long wavelength radio waves. But, like the cinema, empty space is dark. In the words of the Dutch physicist Hugo Tetrode 'The sun would not radiate if it were alone in space and no other bodies could absorb its radiation'[5]. Light needs to encounter a material entity - smoke particles, a cinema screen, the Moon, an astronaut's hand, an eye - to be realised. As the poet Thom Gunn wrote in *Sunlight*, 'what captures light belongs to what it captures.' The transaction between source and destination that summons light from intervening darkness is a quantum event. Sometimes this is referred to as 'collapsing the wave function'. Physicists wince when the phrase is used, as here, by rascally non-physicists because they know that we civilians don't understand what we mean by it; but then, as far as one can gather, no-one, not even quantum physicists, understands it either. William Bragg thought of it as the steady advance of *now* through time; time which 'coagulates a wavy future into a particle past'[6]. It does no harm to appropriate the concept of the collapsing wave function, as a concise way of recognising the essential role that the act of observation plays in making darkness visible. Max Planck, the father of the quantum, went so far as to state 'Es gibt keine Materie an sich' (there is no matter as such)[7]. In the quantum world, to find something out is to cause that something to happen. Quantum coagulation at the point of now accounts for the forms, colours, textures, dimensions, the very essence of reality.

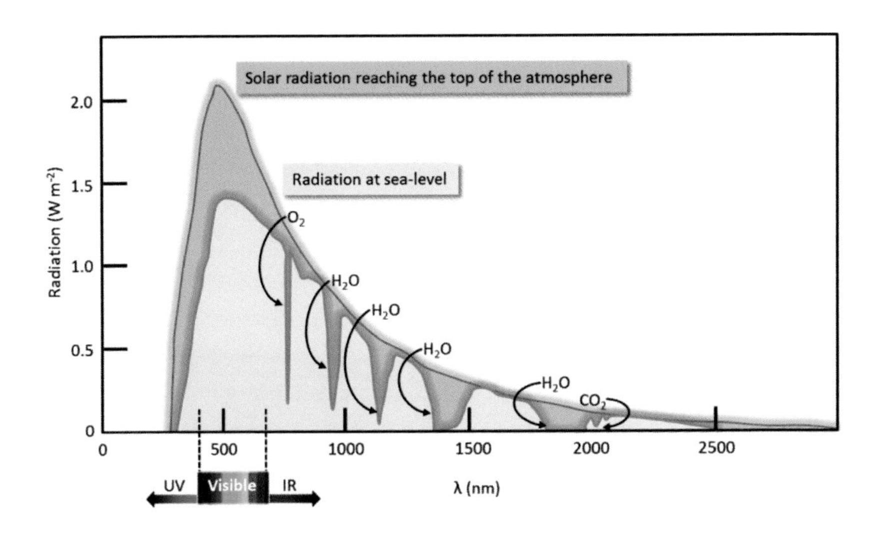

The solar spectrum reaching the upper atmosphere and surface of our planet. Discrete bands in the infrared region of solar radiation are absorbed by atmospheric oxygen, water vapour and carbon dioxide[8]

With the exception of the glimmers of stars and the photons derived from nuclear energy, the light energy that we experience and that shapes our world is entirely solar in origin. The unit of energy is the joule, named after the physicist and brewer James Prescott Joule (Joule's Brewery survives as a business to this day)[9]. The formal definition of one joule is the energy transferred to an object of one kilogram mass moving a distance of one metre when subject to a force required to accelerate it at the rate of one metre per second squared (phew!). In more human-friendly terms, the apple falling from Isaac Newton's tree under the influence of gravity transferred about one joule of energy when it hit the ground. The total annual solar energy absorbed by the Earth's atmosphere, oceans and landmasses is a very large quantity – around 5.62 septillion joules (a septillion is 10^{24} that is, 1 followed by 24 zeroes). Biologists generally define the range and constitution of the electromagnetic spectrum in terms of wavelength (λ, Greek lambda), expressed in units of nanometres (nm; a nanometre is a billionth of a metre) or micrometres (μm; a micrometre is a thousand nanometres). The visible spectrum, which humans see as the colours of the rainbow, runs from a λ of about 380 nm (violet) through to 760 nm (far-red). The energy of a photon is commonly expressed in units of electron-volts. One electron volt is a truly miniscule amount of energy, equal to 16 quintillionths of a joule (a quintillionth is 10^{-18} that is, 0.000000000000000001 - eighteen zeroes after the decimal point). The world of light and energy is indeed one of very large and very small numbers. The energy of a photon is inversely related to its wavelength. Thus gamma- and X-rays, photons with exceptionally short wavelengths, are highly energetic (energies greater than 10 thousand electron volts). At the other end of the electromagnetic spectrum are radio waves, which have extremely long wavelengths and comparatively low energies (less than a millionth of an electron volt). The photon energy of visible light at the short wavelength (blue-violet) end of the spectrum is higher than that at long wavelengths (red) – about 3.3 compared with 1.6 electron volts[10].

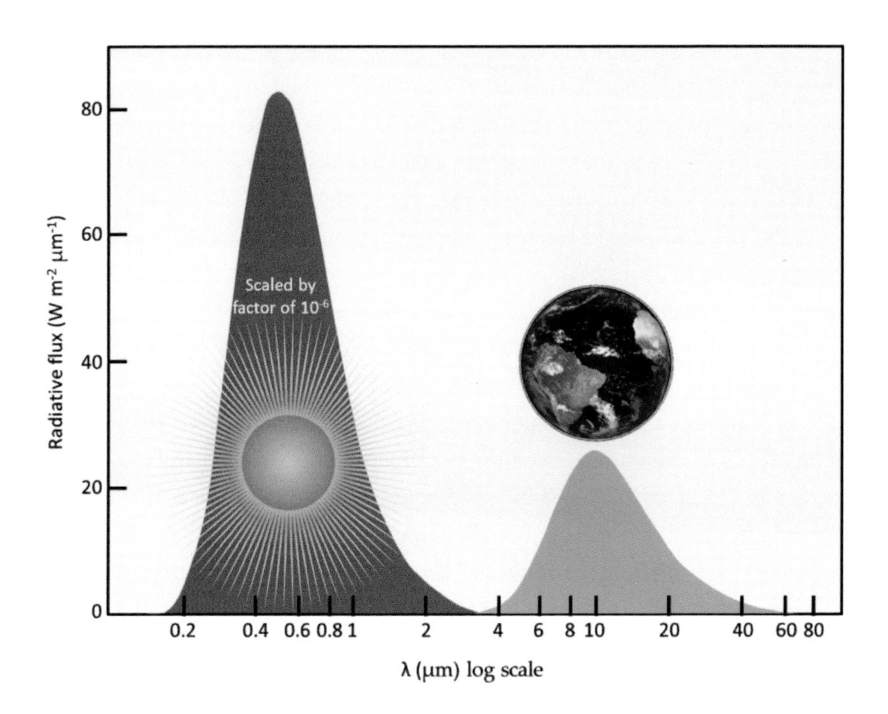

Black body emission curves of the Earth and the sun[11]

A black body is an idealised object or system that absorbs all the energy falling on it, and which maintains thermal equilibrium by emitting radiation at the same rate as it is received. The relationship between a body's temperature and the wavelength of radiation it emits was established by the German physicist Wilhelm Wien. Max Planck regarded 'the spectral density of black body radiation' as something absolute, 'and since the search for the absolutes has always appeared to me to be the highest form of research, I applied myself vigorously to its solution'[12]. Planck's quantum theory allows the distribution of radiant energy across a continuous range of wavelengths to be calculated for a black body of any particular temperature. The sun behaves as a near perfect black body with a surface temperature of 5505 degrees Centigrade. The curve of electromagnetic radiation from the sun arriving at the Earth's outer atmosphere extends from far ultraviolet (UV) at the short wavelength end (about 100 nm or 0.1 μm) into the infrared (IR) with wavelengths of more than 2000 nm (2 μm). The curve peaks in the yellow-green region of the visible spectrum, at around 500 nm. It's no coincidence that this is the wavelength at which the colour sensitivity of human vision is maximal. Much of the energy in the ultraviolet region of sunlight is absorbed by the Earth's ozone layer. The planet's albedo (reflectance, from clouds, snow, ice for example) radiates about 30% of incoming solar energy back into space. The 20% or so of sunlight absorbed by the atmosphere powers the weather, and the 50% that makes it to the earth's surface is what drives biology and geochemistry. Like the sun, planet Earth too behaves as a classical black body. The spectral range of earthlight covers the far infrared, from about 3 to 80 μm with a peak at around 10 μm, corresponding to a temperature of 15 degrees Centigrade. Again, it's no coincidence that this is around the centre of the range of temperatures at which life is possible. The shift towards longer wavelengths in photon energy distribution between incoming sunlight and outgoing earthlight is an example of the second law of thermodynamics in action, which states that entropy (disorder; the amount of thermal energy in a system unavailable for doing useful work) tends to increase. Life on Earth is made possible by solar energy passing through the biosphere: in the memorable words of Lars Olof Björn: '...the energy flux from the sun sweeps through the living world and flushes the disorder produced there out into space'[13].

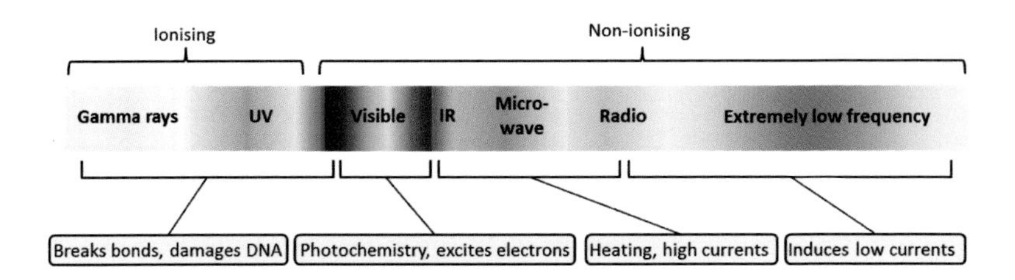

Interactions of electromagnetic radiation with matter[14]

Colour happens when photons meet electrons[15]. Light interacts with atoms in accordance with the principles of quantum physics and thermodynamics. When a photon encounters an atom, it may collide with one of the atom's electrons and transfer its energy to it. An electron, like a photon, is a quantum entity that exists in virtual ambiguity until its wave-function is collapsed. An electron that has acquired energy from a photon is sometimes called an exciton. Short-wavelength photons, including gamma rays, X-rays and far UV radiation, are highly energetic and will give any electron with which they collide enough kinetic energy to kick it out of the atom altogether. By losing a negatively charged electron, the atom becomes a positively charged ion. Short-wavelength ionizing radiation is hazardous to life because of the damage it can inflict on biomolecules. Long-wavelength photons beyond the red end of the visible spectrum aren't energetic enough to make electrons jump; but they can interact with molecules to increase the vibrational energy of their atom-atom bonding systems. It's infrared absorption by carbon dioxide, methane and other greenhouse gases that traps heat in the atmosphere and is a central causative factor in global warming and climate change[16].

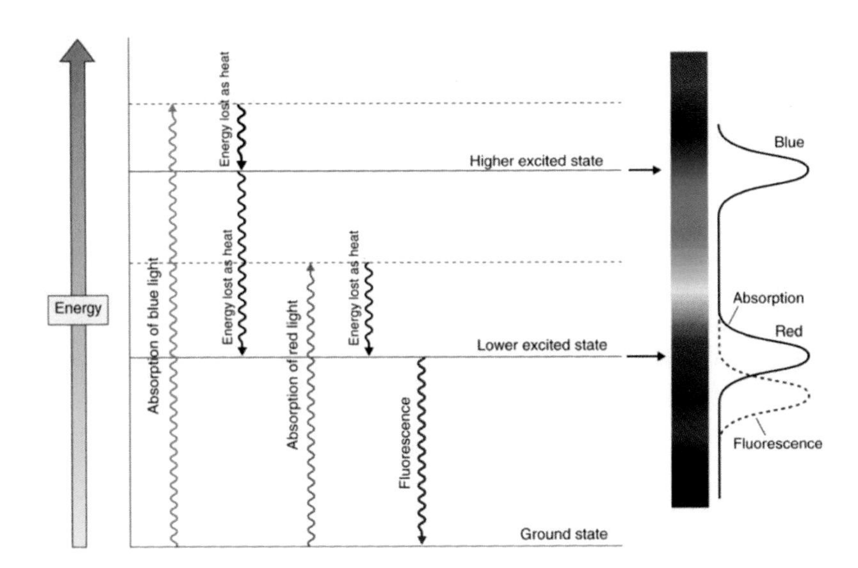

Energy levels within a chlorophyll molecule exposed to visible light[17]

A pigment is a molecule that collapses the wave functions of photons within the non-ionizing visible part of the electromagnetic spectrum, between about 300 and 750 nm. By selectively absorbing certain parts of the visible wavelength range, pigments appear coloured to the human eye (which, as described in characteristically clear and rigorous fashion by Richard Feynman in his lecture on the subject[18], is in turn equipped with three kinds of pigmented receptors). A photon in the visible region of the spectrum is not energetic enough to cause ionization but may make an electron jump to a higher energy-level within an atom of a pigment molecule. An exciton within a pigment may have one of a number of possible fates. It may immediately fall back to its original energy level, simultaneously re-emitting a photon. This is fluorescence. If the excited electron remains highly energised for a relatively prolonged period before giving up its energy and reverting to the low-energy state, it exhibits phosphorescence. Or the exciton may be transferred to another molecule leaving the donor pigment molecule positively charged and the acceptor negatively charged. This is charge separation, and is the process by which photon energy enters the biosphere during photosynthesis[19]. In all cases of pigment-photon interaction, the non-negotiable laws of thermodynamics apply: when an exciton reverts to the low energy state, entropy levels of the system increase and the wavelength of the re-emitted photon is longer (and hence of lower energy) than that of the incoming photon.

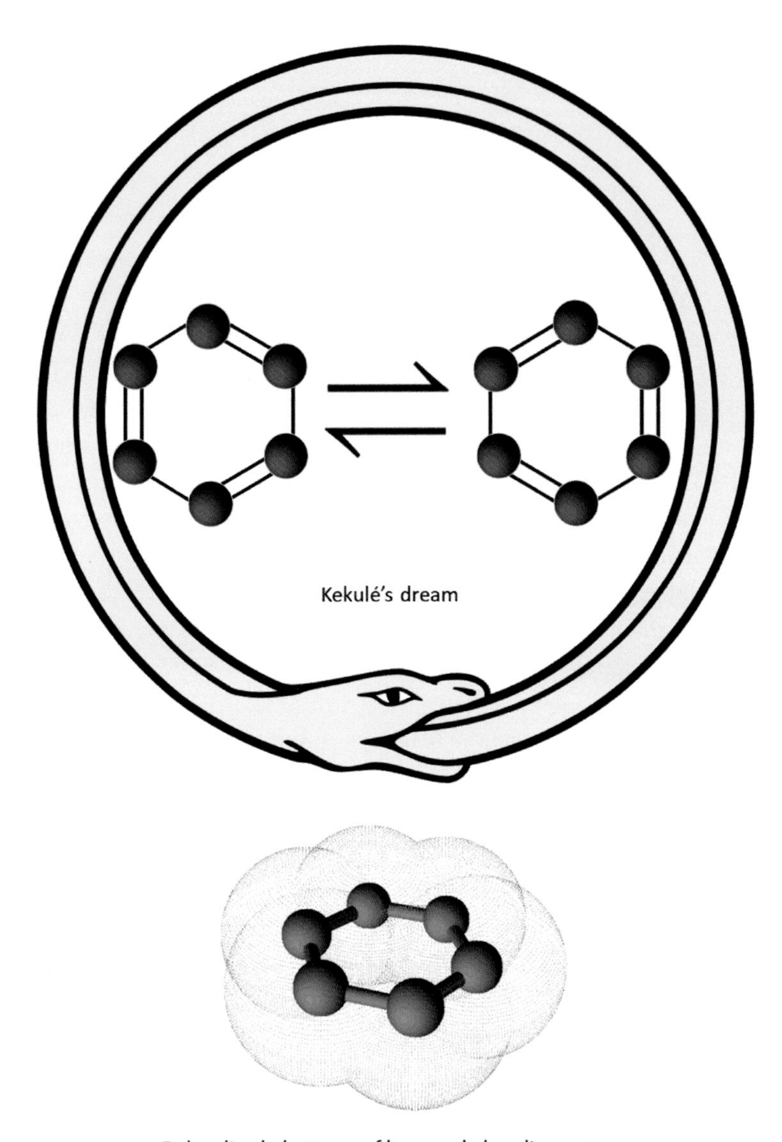

Delocalised electrons of benzene’s bonding system

Structure of the benzene molecule[20]

Living cells are made from a relatively narrow range of different chemical elements. Carbon (symbol C), hydrogen (H) and oxygen (O) are the main constituents, followed by nitrogen (N), phosphorus (P), a few metals and a number of minor elements. Pigment molecules consist of C, H and O plus, in the case of chlorophylls and haems, magnesium (Mg), iron (Fe) and N. P, in the form of the universal energy currency of living cells ATP (adenosine triphosphate), is essential for pigment metabolism. Biomolecules are organic: they have skeletons made from chains, branched structures and/or rings of carbon atoms. A carbon atom has a valence of 4, that is, it can form four bonds with up to four further atoms, including other carbon atoms. Organic chemistry is the chemistry of the four bonds of carbon and the formation of chains and rings with other atoms[21]. One of the pioneers of chemical bonding theory was the brilliant German chemist Friedrich August Kekulé. He struggled to understand the structure of the benzene molecule, which was known to consist of 6 carbon and 6 hydrogen atoms. How could such a composition be reconciled with the known valences of carbon (4) and hydrogen (1)? The (possibly apocryphal[22]) story is told of how the answer came in a dream, in which Kekulé imagined Ouroboros, the ancient symbol represented by the serpent eating its own tail, mutating into a closed ring consisting of 6 carbon atoms, each attached to a single hydrogen atom and joined to one another by alternating single and double bonds. This structural motif, of carbon atoms linked by a sequence of single and double bonds, is called a conjugated bond system and is a recurrent feature of molecules that interact with light within the waveband of the visible and near-visible spectrum. Molecular structures are represented symbolically in different ways, including as chemical formulas, abbreviations of chemical names, two- and three-dimensional structural images and computer-generated models. There will be a fair number of chemical structures later in this book[23], but be not afraid: even if they don't help make the chemistry any clearer, they can be enjoyed as representations of the beautiful molecular sculptures that form the scaffolding on which reality is built.

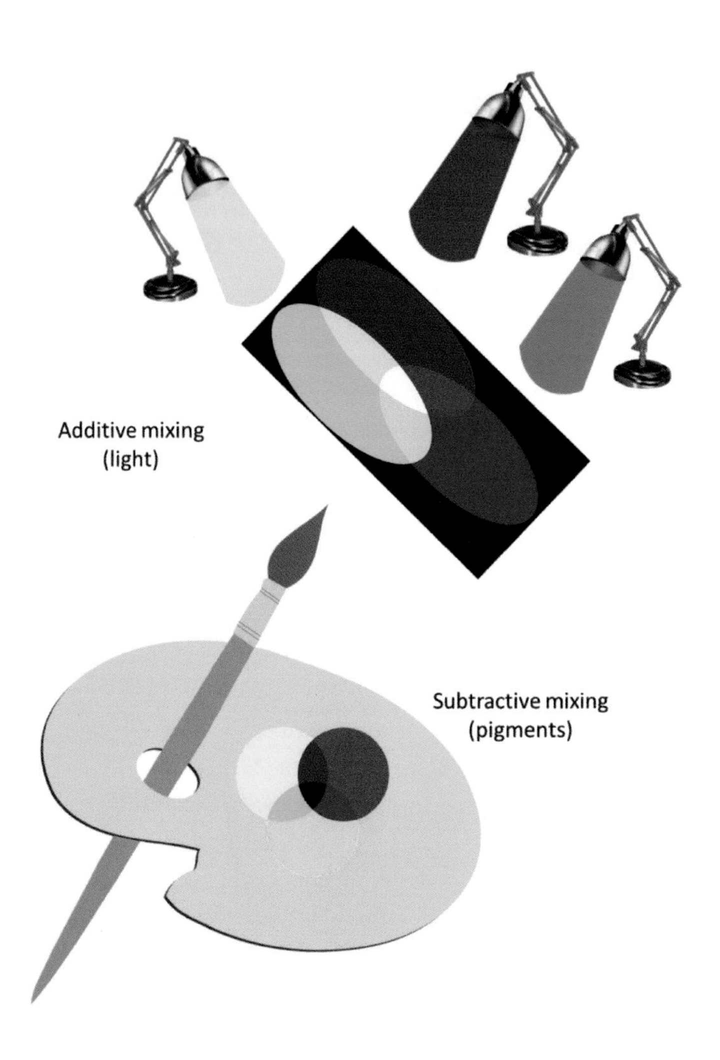

Colours can be combined by addition or subtraction[24]

A pigment molecule will comprise a diversity of atoms in a multiplicity of bonding arrangements, and a corresponding variety of electron ground and excited states. The combined contributions of excitons in this molecular environment define the absorption of light across broad wavelength ranges rather than the sharp bands characteristic of individual atoms: for example, the absorbance spectrum of the green pigment chlorophyll comprises one waveband in the blue and one in the red region. The relationship between reflectance, the visible colour of a pigment, and the colours of light it absorbs is reciprocal. In the case of photosynthesis, the energy required to convert carbon dioxide into organic biomolecules is provided by the light absorbed by chlorophyll; the pigment appears to be green because there is little interaction with photons in the green wavelength range of photon energies. When different pigments are combined, the perceived colour of the mixture is the outcome of subtractive interaction between their respective absorbance and transmittance spectra. Different coloured lights, on the other hand, mix additively (as they do in the digital TV screen, the retinas of our eyes and the outputs of inkjet and laser printers). The distinction between subtractive and additive mixing is the basis of Colour Theory, which is built on critical insights from among the greatest names in science (Isaac Newton, James Clerk Maxwell), the hostility of some other equally great thinkers (notably Johann Wolfgang von Goethe) and even the bafflement of colour professionals such as George Field, supplier of pigments to JMW Turner.

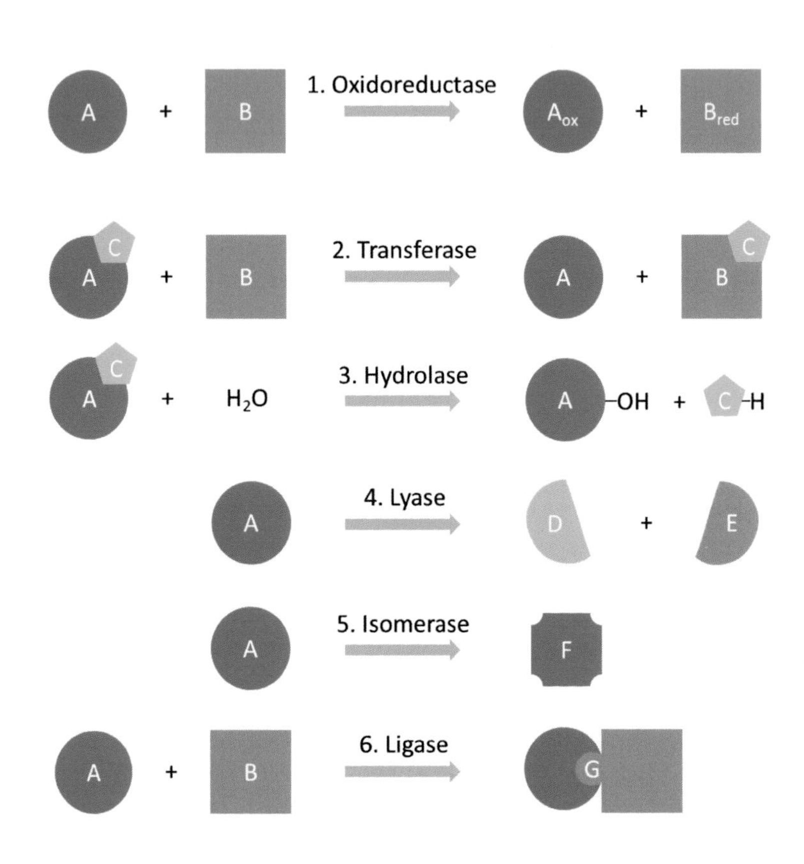

The 6 classes of enzyme[25]

Cells build and dismantle pigments, like all biomolecules, by making and breaking chemical bonds within and between organic molecules (metabolism). Metabolism underlies the enormous diversity of colours, and colour transitions, that adorn the biological world. Chemical changes within living organisms do not happen spontaneously. Each biochemical reaction is facilitated by one or more enzymes, catalysts that interact with reactants thereby overcoming thermodynamic barriers to the formation of products. Broadly speaking there are six different types of enzymic reaction employed by cells to synthesise and break down biomolecules, including pigments. Type 1 comprises redox reactions, in which one molecule, the electron donor, gets oxidised while another accepts the electron and is reduced. Type 2 reactions are the transferases, in which a chemical group is moved from one reactant to another. The transferases include the enzymes that use the energy currency of the cell, ATP, to activate a co-substrate by transferring one of ATP's phosphate groups. Type 3 enzymes are the hydrolases, which catalyse the reaction between a substrate and water. The enzymes that digest proteins and sugars are hydrolases. Type 4 enzymes are lyases, which catalyse the separation of two parts of a molecule, often with the formation of a double bond between them. Type 5 are isomerases, which change one molecule into a different one with neither the addition nor the loss of any atoms. And Type 6, ligases, join together two molecules, often with the expenditure of ATP. That concludes this crash course in the bare bones of biochemistry[26]: it should be enough to make sense of the structure, origins and fates of the three little pigments that are the subject of this book.

NOTES AND SOURCES

[1] Public domain image under CC0 Creative Commons, from https://pixabay.com/en/halo-sun-sky-light-sunlight-rays-364769/ [accessed 25 May 2018].
[2] Image by Bridget Coila under CC BY-SA from www.flickr.com/photos/bibbit/6282788175 [accessed 25 May 2018].
[3] Einstein A. 1951. Letter to Michele Besso.
[4] Public domain image under CC0 Creative Commons, from https://pixabay.com/en/solar-system-planet-space-mars-1789557/ [accessed 25 May 2018].
[5] Tetrode H. 1922. Uber den Wirkungszusammenhang der Welt. Eine Erweiterung der klassischen Dynamik. Zeitschrift für Physik 10: 317–328.
[6] Bragg WL. 1975. *The Development of X-ray Analysis*. London: Bell.
[7] Lecture given in Florence, 1944.
[8] Illustration by author, from various sources. For an account of the relationship between solar radiation and plant colour, see Kiang NY. 2007. The color of life, on Earth and on extrasolar planets. NASA Science Briefs (www.giss.nasa.gov/research/briefs/kiang_01/ [accessed 25 May 2018]).
[9] Cardwell DSL. 1991. *James Joule: A Biography*. Manchester: University Press
[10] Facts and figures about solar energy and the electromagnetic spectrum experienced by the biosphere may be found in chapters 8 and 9 of Jones R, Ougham H, Thomas H, Waaland S. 2012. *The Molecular Life of Plants*. Chichester: Wiley-ASPB.
[11] Illustration by the author, based on various sources. For further information, see www.acs.org/content/acs/en/climatescience/energybalance.html [accessed 25 May 2018].
[12] Sturge MD. 2003. *Statistical and Thermal Physics*. CRC Press, p 201.
[13] Björn LO. 1976. *Light and Life*. Hodder and Stoughton.
[14] Illustration by the author, from various sources.
[15] Jones et al. (2012).
[16] Zhong W, Haigh JD. 2013. The greenhouse effect and carbon dioxide. Weather 68: 100-105.
[17] Illustration from Jones et al. (2012), reproduced by kind permission of the publisher, Wiley.
[18] Feynman R, Leighton R, Sands M. 1963, 2013. *The Feynman Lectures on Physics*. Lecture 35 – Color vision (www.feynmanlectures.caltech.edu/I_35.html [accessed 25 May 2018]).
[19] Dekker JP, Van Grondelle R. 2000. Primary charge separation in photosystem II. Photosynthesis Research 63: 195-208.
[20] Illustration by author, using public domain (CC0) image of the serpent Ouroboros (https://commons.wikimedia.org/wiki/File:Ouroboros-simple.svg [accessed 29 May 2018]).
[21] See chapter 2 of Jones et al. (2012).
[22] Seltzer RJ. 1985. Influence of Kekulé dream on benzene structure disputed. Chemical and Engineering News 63: 22-23.
[23] I've used many sources for the molecular structures presented in these pages. One of the most comprehensive chemical databases is Chemspider (www.chemspider.com/ [accessed 29 May 2018]). ChEBI is another useful source (www.ebi.ac.uk/chebi/). Most of the original

and modified structures were created in Chemsketch (www.acdlabs.com). These are excellent resources, without which illustrating this book would have been entailed much weeping, wailing and gnashing of teeth.
[24] Ball P. 2008. *Bright Earth: the Invention of Colour*. London: Vintage.
[25] Enzyme classification was formalised in 1961 (with many subsequent supplements) by the Enzyme Commission (EC) of the International Union of Biochemistry and Molecular Biology (www.sbcs.qmul.ac.uk/iubmb/enzyme/ [accessed 29 May 2018]). BRENDA is a compendious list of enzymes with their EC numbers (www.brenda-enzymes.org/all_enzymes.php [accessed 29 May 2018]). The present book draws extensively on KEGG (www.genome.jp/kegg/kegg2.html [accessed 29 May 2018]), a comprehensive database of the pathways of enzyme-catalysed metabolism.
[26] Chapter 2 of Jones et al. (2012) is a recommended text (it ought to be relevant, since I wrote it…)

2. GREEN

Concerning chlorophylls and their blue and red relatives[27]

Why grasse is greene, or why our blood is red,
Are mysteries which none haue reach'd vnto.

John Donne (*The Second Anniversarie of The Progres of the Soule*)

The Greene Lyon eats the sun[28]

Isaac Newton was a man with one foot in modernity and one in the medieval world: in the words of Betty Jo Teeter Dobbs, his was 'the Janus face of genius'. With what John Keats dismissed as 'the mere touch of cold philosophy', Newton ushered in the modern age of colour theory. But while he struggled with 'the dull catalogue of common things', Newton was living a semi-clandestine life as an alchemical mystic. At his death in 1727, many of the manuscripts he left, mostly in his own hand, were considered unfit to be printed because their contents were at odds with his exalted status as the founder of classical physics. Arguably it was the auction of a tranche of Newtoniana by Sotheby's in 1936 that brought the occult obsessions of the great man to light. Among allusions to the mysterious and graphic imagery of The Craft, Newton repeatedly refers to the 'Greene Lyon'[29]. 'The Hunting of the Greene Lyon' is an ancient alchemical verse which includes the lines '...yet full quickly he can run,/ And soone can overtake the Sun:/ And suddainely can hym devoure...' There is good reason to see the influence of 'corpuscular alchemy' in Newton's notebooks at the time when he was developing his theory of colour[30]. Though it's usually understood to be descriptive of colour change during elemental transmutation, I have often wondered whether the image of the sun being consumed by a green living thing might represent an early esoteric insight into photosynthesis. It would not be the first time that the alchemical world-view anticipated modern physico-chemical science. It was not until fifty years after Newton's death that Jan Ingenhousz showed the light-dependence of plant growth and oxygen evolution.

Starch print of Julius von Sachs. A leaf exposed to light through a photographic negative makes starch where light gets through. Starch forms a dark purple complex when the tissue is treated with iodine.

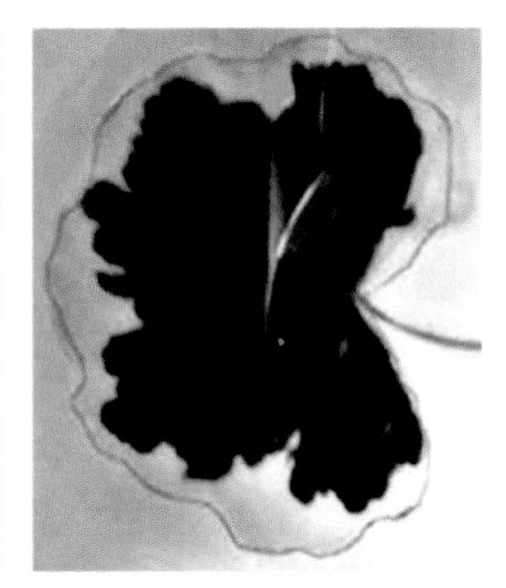

A variegated leaf exposed to light makes starch only where there's chlorophyll

Starch prints[31] show that photosynthesis needs light and chlorophyll

The French chemists Pierre-Joseph Pelletier and Joseph-Bienaimé Caventou were the first to isolate 'chlorophyle', (from the classical Greek *khloros*, green and *phyllon*, leaf) in 1817. It was assumed that chlorophyll was uniformly distributed within the cells of green tissues, until the German physiologist Julius von Sachs showed, in 1883, that the pigment is located in the subcellular structures we now call chloroplasts. He also raised the possibility that the function of chlorophyll is to capture the light which Ingenhousz had shown is necessary for photosynthesis. There are two forms of chlorophyll in land plants, a and b. The German chemist Richard Willstätter succeeded in separating and isolating chlorophyll a and chlorophyll b, and received the Nobel prize for this work in 1915. Determining the chemical structure of chlorophyll molecules is challenging and it was not until 1940 that the German Nobel laureate Hans Fischer succeeded. It took a further decade before the outline of the biochemical mechanism by which chlorophyll was biosynthesised became clear, and almost half a century elapsed before the disposal pathway for chlorophyll was discovered.

Solar panels[32]

Before we go into the subjects of the colour, chemistry and function of chlorophyll, we should try to answer the sort of naively profound question a child might ask: why, if leaves evolved to intercept light energy and therefore to act as solar panels, are they green and not black? Here's a possible explanation for the origin of leaf colour. It's better for a solar panel to be black because it absorbs light more or less equally from the blue to red end of the visible spectrum. Some people think that, early in the evolution of life on Earth, the pigments of the first photosynthetic lifeforms did indeed absorb light across most of the middle range of the spectrum[33]. Organisms that take light over these wavelengths would appear to the eye as dark purple in colour – not so far from the solar panel ideal. By monopolising the central wavebands, these early purple lifeforms would have created a niche: if an organism evolved with pigments that could use light at the extreme red and blue ends of the spectrum, it could make a living. Chlorophyll has just such a light absorption profile and appears green to the eye. For some reason, the green photosynthesisers prevailed during evolution and became the ancestors of all the vegetation that surrounds us today. Meanwhile the purple types never diversified beyond the single-celled condition and their descendants, the purple bacteria, are now confined to extreme marginal habitats such as hot sulphurous springs[34].

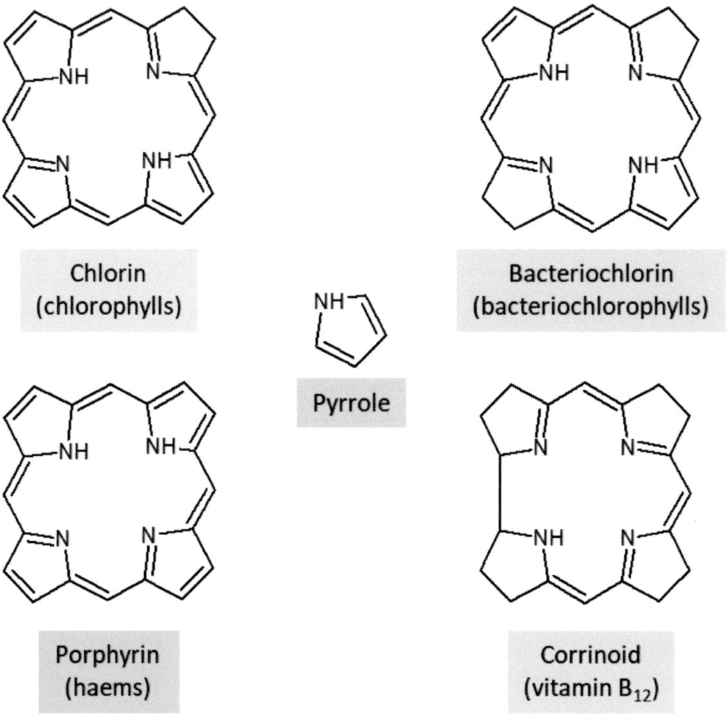
NH
N
N
NH
Chlorin
(chlorophylls)
NH
Pyrrole
NH
N
N
NH
Bacteriochlorin
(bacteriochlorophylls)
NH
NH
N
N
Porphyrin
(haems)
N
N
NH
N
Corrinoid
(vitamin B_{12})

Core structures of tetrapyrrole macrocycles[35]

The chlorophylls are a family of pigments with a common chemical structure based on a macrocycle; that is, a ring made up of a number of smaller rings. Each of these component rings is a 5-sided structure containing a nitrogen atom (a pyrrole group). The ring of four pyrrole groups is named, not surprisingly, a cyclic tetrapyrrole. Curiously, free pyrrole itself is not naturally found in living cells (although it's a component of smoke and coal tar), and neither it nor its close chemical relations occurs as an intermediate in the metabolism of chlorophylls and similar biomolecules (though it has been used as a starting material for chemical syntheses of tetrapyrroles in the laboratory). In a tour-de-force of synthetic chemistry, the group led by Robert Woodward (who memorably called chlorophyll 'the green badge of life') finally succeeded in building the entire chlorophyll molecule de novo[36], an achievement that gained Woodward the Nobel Prize for Chemistry in 1965. Several variants of tetrapyrrole structure are known, distinguished by the extent of double-bonding around the macrocycle. The chlorophylls of green plants are derivatives of chlorin. The haem pigments, which include the component of haemoglobin that gives blood its red colour, are based on the tetrapyrrole porphyrin (incidentally, this book uses the British spelling haem rather than US form heme, though the latter will get many more hits in an online search). And vitamin B_{12} is a derivative of corrin, a representative of the group of tetrapyrroles called corrinoids. Vitamin B_{12} was one of a number of important biomolecules that Robert Woodward's group chemically synthesised as well as chlorophyll[37].

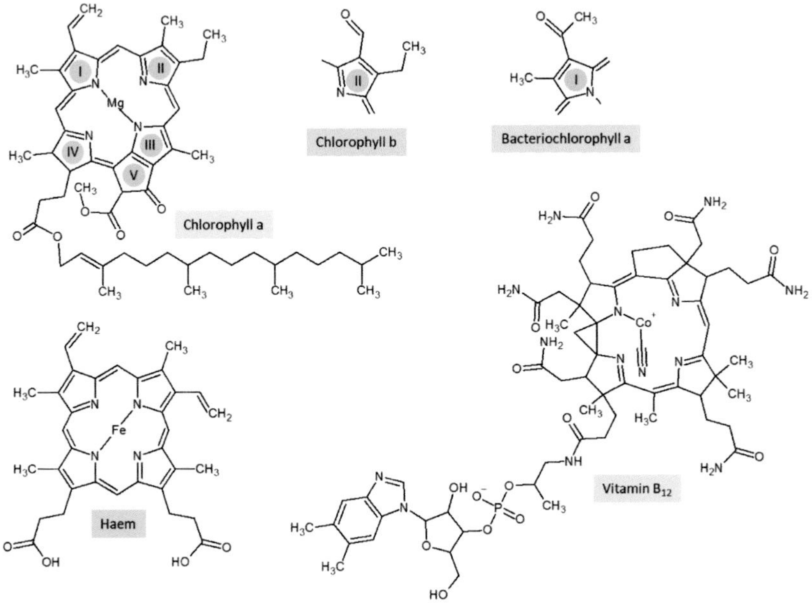

Structures of tetrapyrrole pigments and vitamin B_{12}[38]

Living cells seem to have acquired the capacity to make and use tetrapyrroles very early in biological evolution, but the details are obscure. According to Taniguchi, Ptaszek, Chandrashaker and Lindsey[39] 'the questions of whether tetrapyrrole macrocycles formed in the prebiotic era and may have contributed to the origin of life are not known and indeed may not be answerable with scientific certainty.' Attempts have been made to recreate plausible early non-enzymic mechanisms leading to products that became progressively refined and replaced as evolution proceeded. But the upshot is that tetrapyrrole biochemistry in extant organisms is, to use the striking term of Taniguchi and co-authors[40], a palimpsest - a later record that has overwritten and completely obscured the original. It took 55 separate chemical reactions, many of them novel and technically challenging, for Robert Woodward's group to achieve the total chemical synthesis of chlorophyll (in another of his striking phrases, Woodward wrote of entering a 'chemical fairyland'). Plants are able to accomplish the same feat in about 20 biochemical steps. At least four of these enzyme reactions require oxygen. The atmosphere of the early Earth was effectively anaerobic, until the Great Oxidation Event (GOE), a rapid increase in atmospheric oxygen concentration that occurred around 2.5 billion years ago[41]. There are at least three scenarios for the appearance and evolution of tetrapyrrole biosynthesis in the period before and after the GOE[42]. All in all, it seems likely that tetrapyrroles have been around for perhaps 3 billion years. The tetrapyrrole macrocycle, with its four inward-pointing nitrogen atoms, makes a kind of nest, within which metal atoms can be held. Cells clearly find such structures to be very handy for all manner of biochemical jobs[43]. In the case of haem, an iron atom (Fe) occupies the middle of a porphyrin-derived tetrapyrrole. In vitamin B_{12}, the nitrogens of the corrin ring complex with cobalt (Co). And in chlorophylls and bacteriochlorophylls, the metal atom in the chlorin macrocycle is magnesium (Mg). An array of chemical groups may be attached to the ring of rings[44].

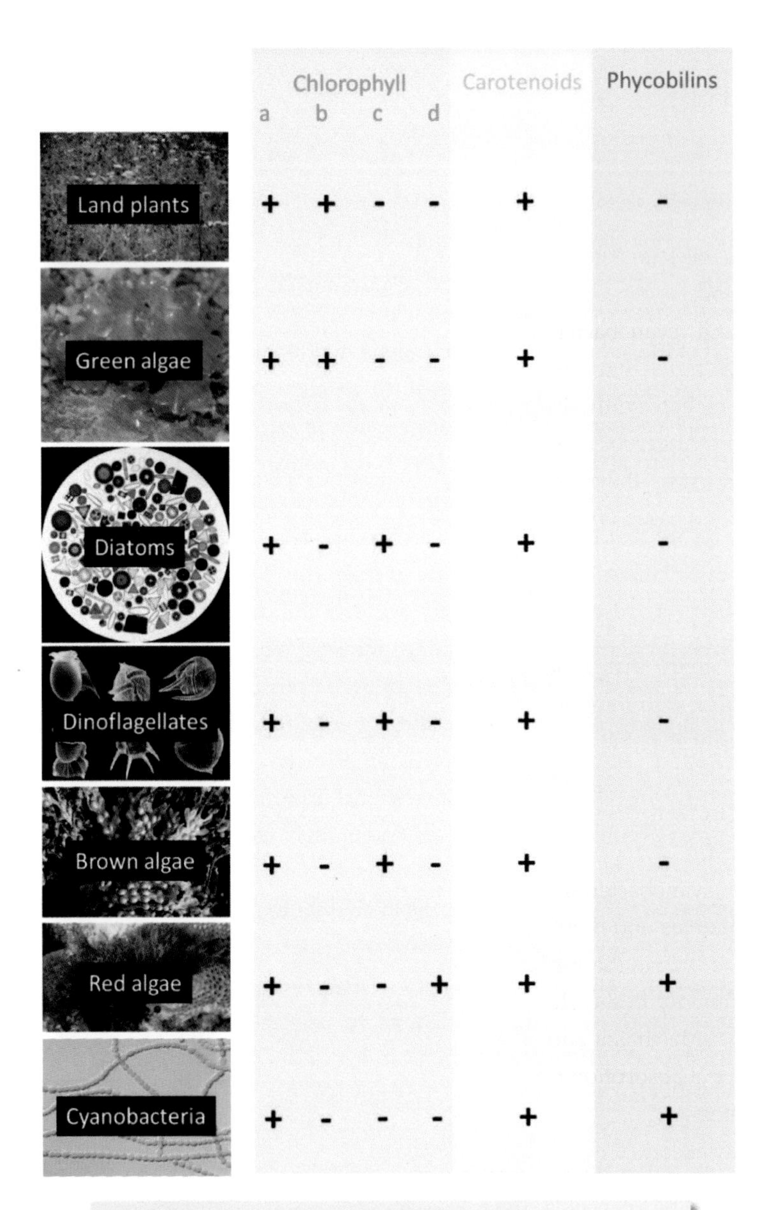

	Chlorophyll a	Chlorophyll b	Chlorophyll c	Chlorophyll d	Carotenoids	Phycobilins
Land plants	+	+	-	-	+	-
Green algae	+	+	-	-	+	-
Diatoms	+	-	+	-	+	-
Dinoflagellates	+	-	+	-	+	-
Brown algae	+	-	+	-	+	-
Red algae	+	-	-	+	+	+
Cyanobacteria	+	-	-	-	+	+

Pigments of plants capable of oxygenic photosynthesis[45]

Let's take a look at the different kinds and colours of chlorophyll[46]. Chlorophyll is the primary pigment absorbing light energy in the plants, algae and cyanobacteria that carry out oxygenic photosynthesis, that is, photosynthesis that produces oxygen by splitting water (the appearance of oxygenic photosynthesis in early evolution is believed to have been the cause of the GOE). Anaerobic photosynthetic bacteria tend to occupy red-depleted aquatic environments; they absorb light through bacteriochlorophyll, a form of chlorophyll based on the bacteriochlorin macrocycle. Bacteriochlorophyll extends the waveband of useable light beyond 700 nm into the less energetic wavelengths of the near infrared. The tetrapyrrole ring of chlorophyll and bacteriochlorophyll, which binds a central magnesium atom, is attached to phytol, a long chain of 18 carbon atoms that renders the molecule extremely insoluble in water but compatible with the lipids of chloroplast membranes. Chlorophylls a, b, c and d are variants on the basic chlorophyll structure, differing in the nature of the chemical side groups on the macrocycle. All oxygen-evolving photosynthetic organisms contain chlorophyll a, and in almost all cyanobacteria chlorophyll a is the sole form of chlorophyll. Green land plants, green algae, and one or two unusual cyanobacterial species, contain a second form, chlorophyll b. Brown algae, diatoms and dinoflagellates contain chlorophyll c as well as chlorophyll a, and in red algae the forms of chlorophyll are a and d. All chlorophylls absorb light primarily in the blue and red wavelengths of the visible spectrum, but the different side-chains and bonding patterns of the macrocycle significantly alter the spectral absorption profiles of the various chlorophyll species. The colour of the photosynthetic tissues of land plants is due to reflection of green light, which is weakly absorbed by chlorophylls a and b.

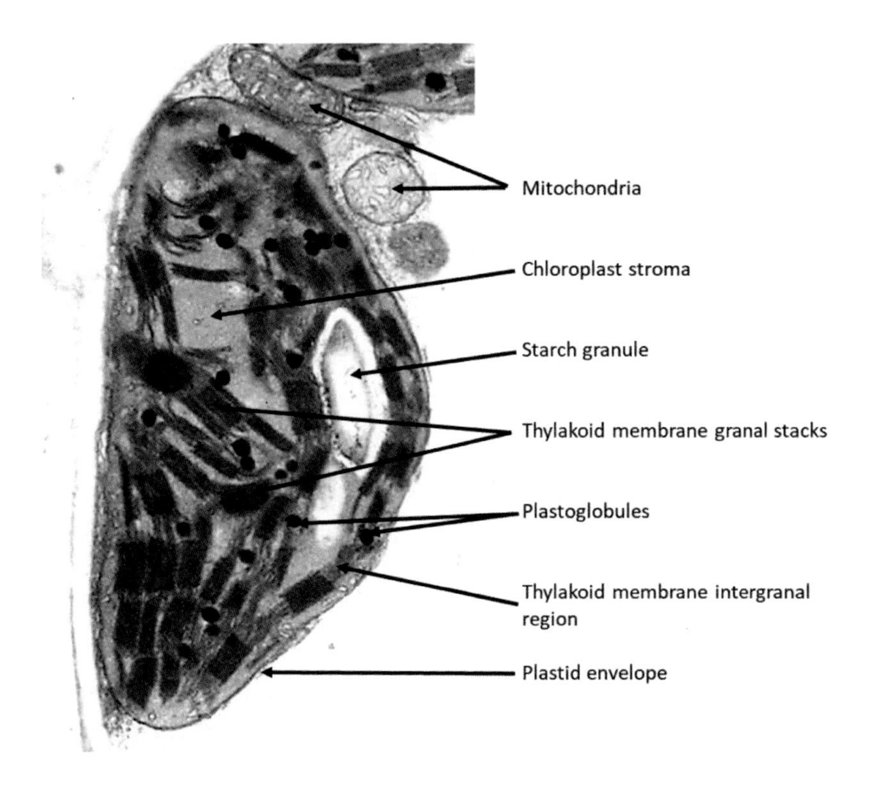

Electron micrograph of a chloroplast (dimensions 2.1 x 4.8 μm) from a green leaf of *Lolium temulentum*[47]

Photosynthesis is how solar energy and the carbon for biochemistry enter the biosphere. For light energy to have a biological effect, it must be intercepted by a pigment. The chlorophylls perform this function in photosynthesis. The characteristic organelles of plant cells are plastids, first recognised and described in the mid 19th century by the German biologists Ernst Haeckel and Andreas Schimper[48]. Chlorophyll is located in chloroplasts, the specialised green photosynthetic representatives of the plastid family within the cells of algae and land plants. A chloroplast consists of a protein-rich stroma matrix within which is embedded an internal system of thylakoid membranes. The thylakoid system of green plants is organized into stacked regions, called grana, with unstacked membranes exposed to the stroma in between. Thylakoids and stroma are surrounded by a double membrane envelope. The photosynthetic pigments and associated proteins are localised in the thylakoids and are organised into two photosystems, PSI and PSII. These two pigment-protein complexes are structurally, functionally and spatially distinct, but share features with all photosystems across the range of photosynthetic organisms. This suggests that photosystems have a common evolutionary origin. The chlorophyll molecule is flattish and the electrons that bond its atoms together exist as a delocalised cloud above and below the plane of the molecule. Chlorophyll molecules group together to form antenna arrays, often called light-harvesting complexes[49], and their delocalised electrons become fused into, effectively, one big shared cloud. This means that when a photon interacts with any electron in the cloud, its energy is transferred at high speed, and with close to 100% efficiency, through the whole antenna array until it reaches a photosynthetic reaction centre where it can be used to release oxygen from water and ultimately to fix carbon dioxide. In most plants, each photosystem complex comprises about 250 antenna chlorophyll a and b molecules associated with a reaction centre.

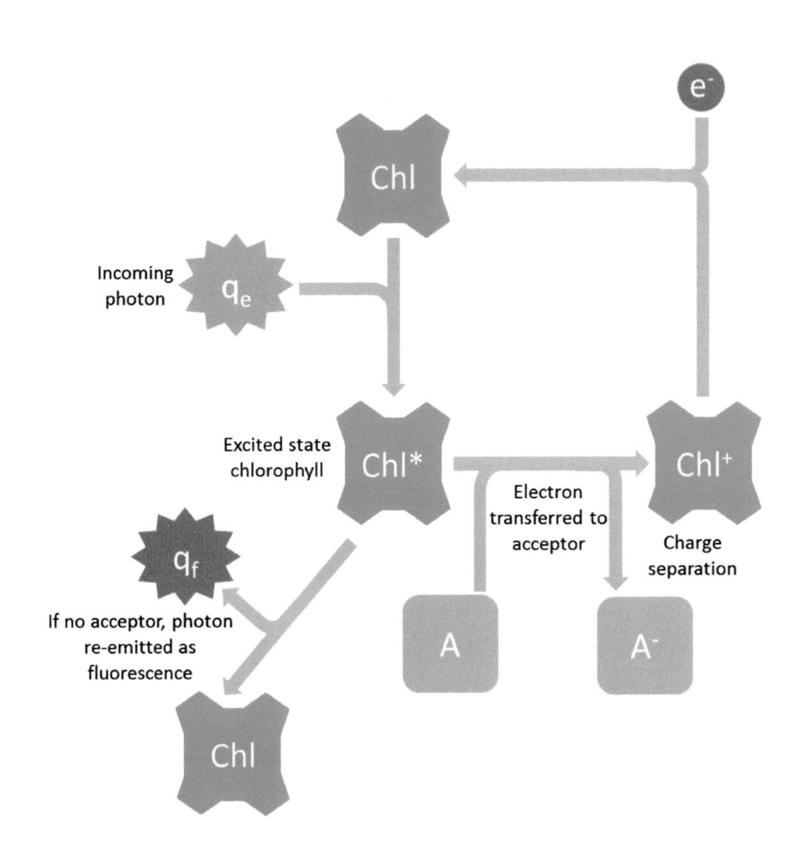

Energy dissipation from excited-state chlorophyll

We've seen that, when a pigment absorbs a light quantum, one of its electrons jumps to an excited state. The excited electron may be transferred to an acceptor molecule in a process called charge separation. Charge separation is the essential photochemical event of photosynthesis and takes place at the reaction centre of each photosystem. The sequence of events may be summarised as:

$$Chl + q_e + A \rightarrow Chl^* + A \rightarrow Chl^+ + A^-$$

where q_e is the exciting photon, A is the acceptor, Chl is chlorophyll and Chl* denotes the excited state of Chl. In the charge separation cycle, Chl^+ is reduced back to Chl by an electron from water (PSII) or by one delivered to PSI by plastocyanin, and the cycle continues. If electron transfer from Chl* to A fails, the pigment may revert to the ground state and will fluoresce - that is, it emits a photon at a longer wavelength than that of the exciting photon (q_f).

$$Chl^* \rightarrow Chl + q_f$$

The rate of transfer of an excited electron from Chl* to an acceptor is about a thousand times that of fluorescence emission. The efficiency of photosynthesis - referred to as quantum yield - can be estimated by measuring chlorophyll fluorescence. Under optimal conditions, with an adequate supply of available acceptors, reaction centre photochemistry can be almost 100% efficient, but physiological and environmental stresses can have the effect of increasing fluorescence and reducing quantum efficiency. Various handy point-and-shoot gadgets have been developed that enable crop scientists, agronomists and even farmers to monitor non-invasively the degree of stress experienced by plants in the field by measuring chlorophyll fluorescence[50]. If for any reason photon energy can't be delivered to a reaction centre, light-excited chlorophyll and derived tetrapyrroles may become hazardous, a source of reactive chemical species that can damage or kill the cell[51]. This is photosensitivity, which is an ever-present threat for plants, whose primary energy source is light, making it necessary for defences against damage by excess light to be built in to plant cell structure and function.

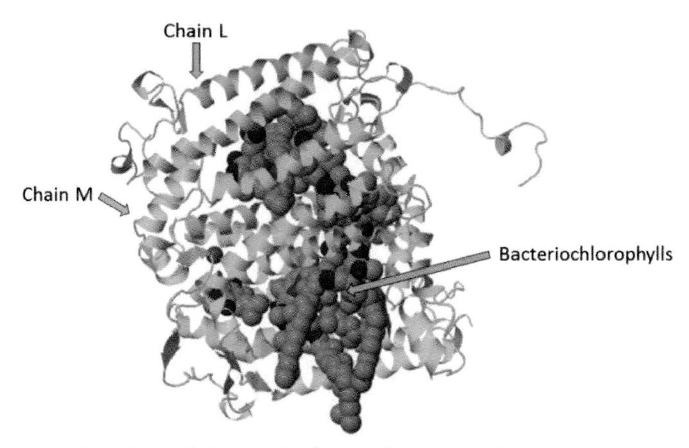

Proteins and pigments in the crystal structure of
the *Rhodopseudomonas* reaction centre

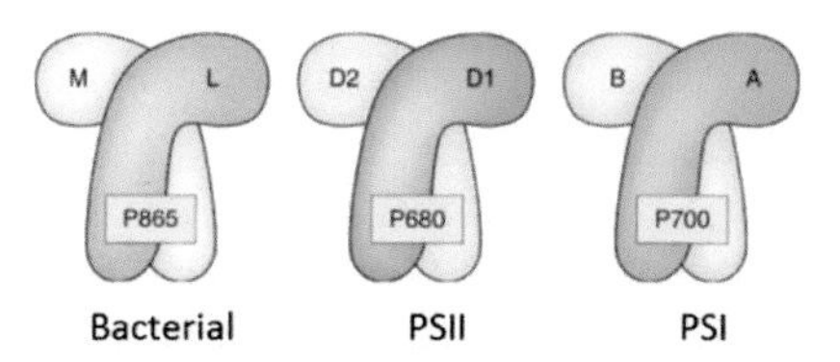

Bacterial and green plant photosynthetic reaction centres compared[52]

We know what a reaction centre looks like, and how it works, in exquisite atomic detail thanks to the pioneering Nobel prize-winning X-ray crystallography research of Johann Deisenhofer, Robert Huber and Hartmut Michel[53]. These studies revealed the molecular structures and interactions of the reaction centre complex isolated from the photosynthetic purple bacterium *Rhodopseudomonas viridis*. The three-dimensional configuration of the *R. viridis* reaction centre has proven to be the conceptual model for the general structures and functions of reaction centres and photosystems from all photosynthetic organisms, including multicellular green plants. At the heart of the *R viridis* reaction centre are two bacteriochlorophyll molecules, known as the special pair. This is where charge separation takes place, and is often referred to as P865, because 865 nm is the characteristic maximum wavelength at which the special pair absorbs light in the reaction centre environment. The pigments are associated with twinned, structurally similar proteins, L and M, arranged symmetrically in the photosynthetic membrane. The reaction centres of multicellular green plants are structurally and functionally homologous to that of the purple bacterium. The PSII reaction centre comprises D1 and D2, proteins equivalent to L and M in *R viridis*, and the dimer of chlorophyll a is referred to as P680 by virtue of its absorption maximum at 680 nm. The comparable structures in PSI are protein subunits A and B and pigment dimer P700. It's thought that the structural redundancy within reaction centres, and the diversity of photosystems between organisms, have arisen by duplication and evolutionary divergence of a basic unitary structure.

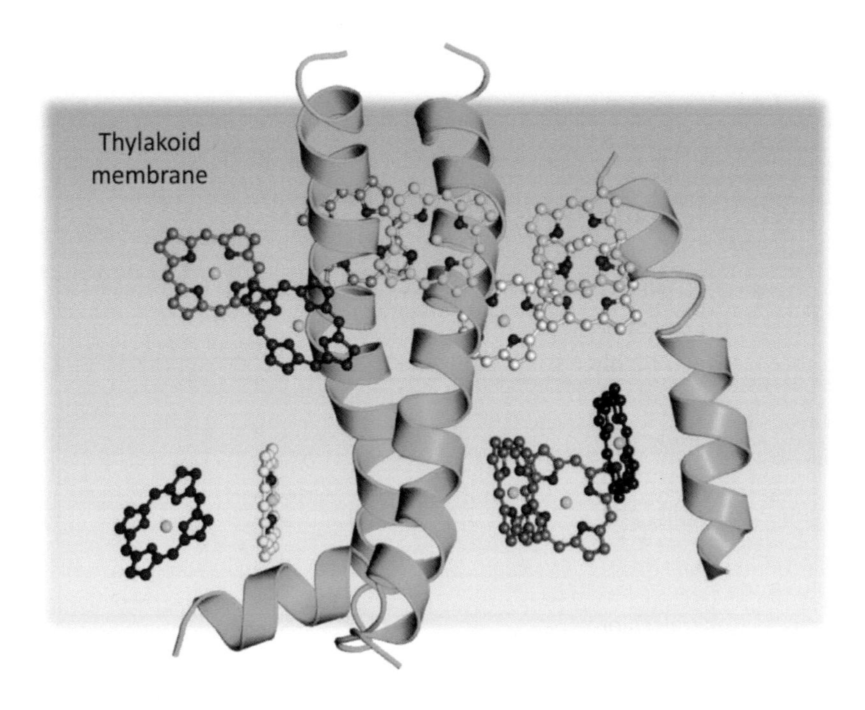

Molecular structure of one of the three subunits of the light harvesting antenna complex of PSII, showing membrane-spanning protein helices and positions of chlorophyll a and b molecules[54]

Plants use chlorophyll to capture the energy of light and to split water molecules. Photons collected by light-harvesting chlorophylls are funnelled to reaction centre chlorophylls where the process of charge separation makes light energy available in the form of the electron energy that powers the biochemistry of photosynthesis. The PSII reaction centre, consisting of P680, D1, D2 and other components, is associated with light-harvesting antennae, formed from combinations of three different chlorophyll-binding proteins, each of which is attached to a dozen or so chlorophyll a and b molecules. The light-harvesting complex of PSII is highly abundant, accounting for most of the green colour of leaves. It also constitutes about a third of leaf protein - for which grazing animals should be grateful. PSII is essentially a photovoltaic device, in which charge separation splits water molecules into electrons and molecular oxygen. All the oxygen we breathe, and just about all the atmospheric oxygen throughout the history of the planet, right back to the GOE, comes from the activity of photosynthetic reaction centres. PSI also has light-harvesting antennae consisting of four peripheral chlorophyll-protein subunits, each associated with about a dozen chlorophyll a and b molecules. When the electrons that come indirectly from the water-splitting activity of PSII arrive at the PSI reaction centre, charge separation drives the formation of reducing power that fixes CO_2 into organic compounds.

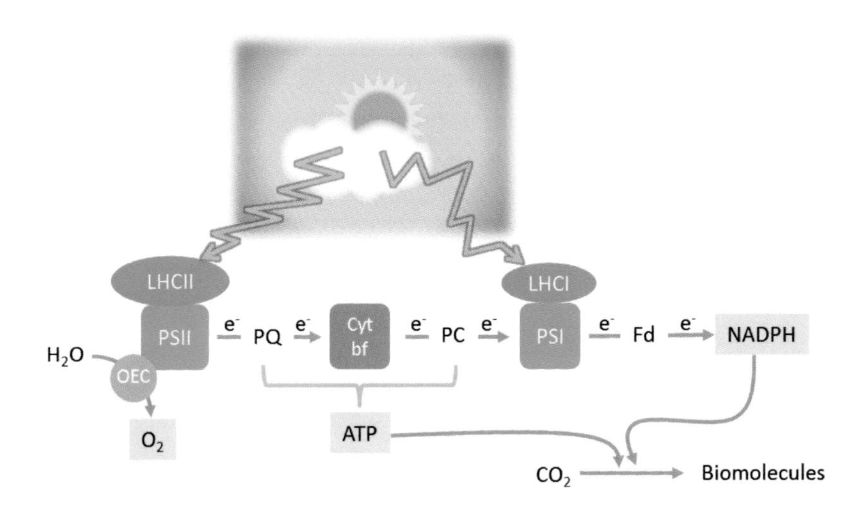

Photosynthesis, from photons to pigments to electrons to biomolecules

We can now build a generalised model showing how plants deploy chlorophyll to drive the biosynthesis of organic molecules from CO_2 and water. The light harvesting antenna complex of PSII (LHCII) traps photon energy which ultimately powers charge separation in the PSII reaction centre. Through the oxygen-evolving complex associated with PSII (OEC), charge separation actuates the release of one molecule of O_2, 4 hydrogen ions and 4 electrons (e^-) from every two water molecules. Electrons are carried to the cytochrome bf complex (Cyt bf) by plastoquinone (PQ), and from Cyt bf to PSI by plastocyanin (PC). Electrons arriving at PSI are given a further boost in the reaction centre by photon energy collected by the light harvesting complex (LHCI). Electrons proceed onward, carried by ferredoxin (Fd), and finally reduce NADP. The energy of electrons passing along the transport chain from PSII to PSI is captured in ATP. Enzymes in the chloroplast use ATP and NADPH to fix CO_2 and form the organic products of photosynthesis.

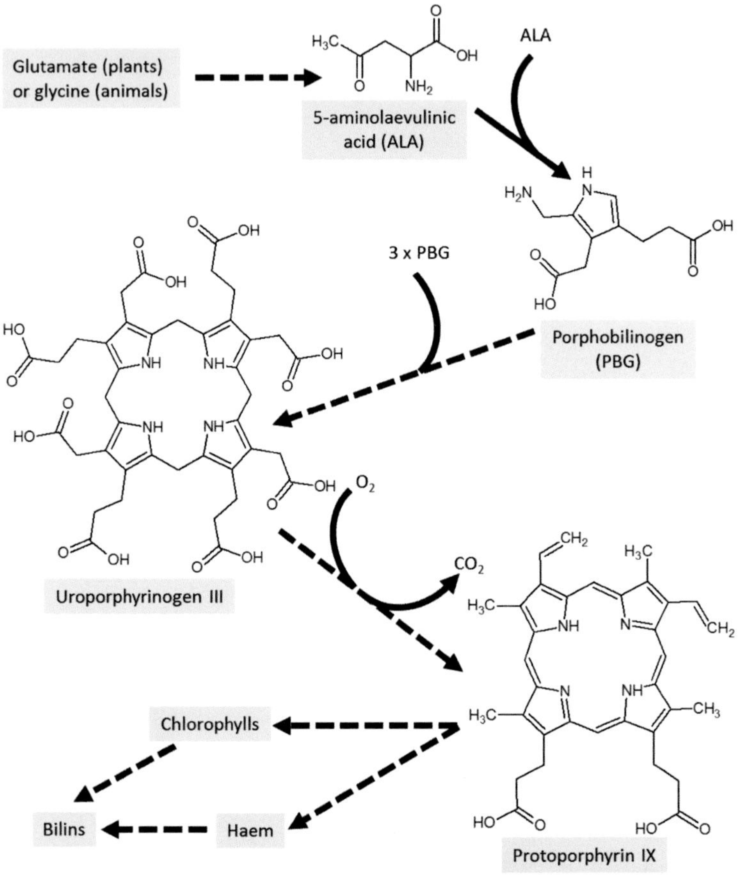

Glutamate (plants) or glycine (animals)

5-aminolaevulinic acid (ALA)

ALA

Porphobilinogen (PBG)

3 x PBG

Uroporphyrinogen III

O_2

CO_2

Protoporphyrin IX

Chlorophylls

Haem

Bilins

Tetrapyrrole biosynthesis as far as the branchpoint

As little pigments go, the chlorophyll molecule isn't so little, and we know from the Woodward chemical synthesis that building it is a complicated business[55]. The scheme presented here shows the bare bones of the first part of the pathway of tetrapyrrole biosynthesis. It has its origins in the core biochemistry of the cell, specifically the amino acids from which proteins and other essential biomolecules are built. The amino acid glycine (in animals and fungi) or glutamate (plants) is converted to the universal first committed intermediate in tetrapyrrole synthesis, ALA. The supply of ALA, a small molecule with a backbone of five carbon atoms to which is attached an amino ($-NH_2$) group, is one of the points at which the overall rate of tetrapyrrole production is regulated. It's under the control of light, and is also feedback-inhibited by haem. In the next step, two ALA molecules are combined to make PBG, and here we see the first appearance of a pyrrole ring structure. In a series of enzyme-mediated steps, four PBG molecules are joined up to form uroporphyrinogen III, the first intermediate with a tetrapyrrole macrocycle. Uroporphyrinogen is festooned with a whole array of carboxyl (-COOH) side-groups, most of which are now removed as CO_2 in a few enzymic steps that require molecular oxygen. And so we arrive at protoporphyrin IX, which can undergo two different metabolic fates[56] and thus stands at a branch-point. If ferrous iron (Fe^{2+}) is inserted into the centre of the protoporphyrin ring the molecule is destined to become haem. Insertion of magnesium (Mg^{2+}) leads to the synthesis of chlorophyll. In plants these processes are carried out by the metal chelatase enzymes ferrochelatase and magnesium chelatase, respectively. We'll look at the iron route presently, but first we'll take a trip along the magnesium pathway.

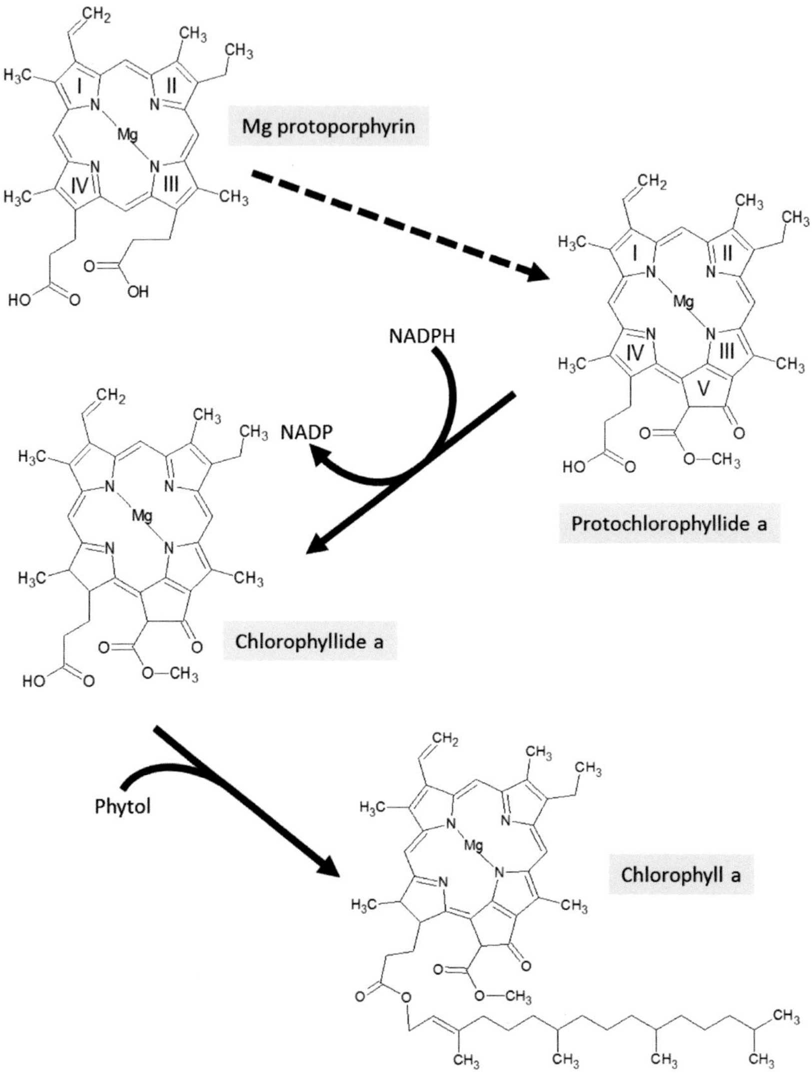

Mg protoporphyrin

Protochlorophyllide a

NADPH

NADP

Chlorophyllide a

Phytol

Chlorophyll a

Biosynthesis of chlorophyll a

Magnesium-protoporphyrin is the first intermediate on the way from the tetrapyrrole synthesis branchpoint to chlorophyll. It's converted to protochlorophyllide a in three enzymic steps, one of which adds a 5-carbon ring (isocyclic ring V) to the four pyrrole rings of the macrocycle - a characteristic structural feature of all chlorophylls. Chlorophyllide a, the immediate precursor of chlorophyll a, is the product of reduction of protochlorophyllide by a remarkable enzyme, NADPH-protochlorophyllide oxidoreductase (POR). LPOR, a form of the enzyme whose catalytic function requires exposure to light, occurs in all photosynthetic organisms (except for a few anaerobic bacteria). DPOR, an additional, light-independent, enzyme (structurally unrelated to LPOR) is present in many algae, aerobic photosynthetic bacteria, liverworts and gymnosperms but, for some mysterious reason, disappeared without trace from flowering plants early in their evolution[57]. Plants possessing DPOR can make chlorophyll in the dark, a capability that flowering plants have lost (it's one of the reasons why, if you cut a green cabbage in half, the inner leaves, which have not been exposed to light, are white). LPOR forms a complex with its substrates and in this state it behaves as a photoreceptor, with the protochlorophyllide pigment attached to the enzyme protein acting as its light-sensor. Absorbed photons trigger the formation and release of chlorophyllide a. The enzyme molecule must immediately bind another protochlorophyllide molecule and repeat the catalytic cycle or else it gets destroyed. The complex relationship of LPOR with its turnover, its substrates and light (which not only facilitates LPOR's catalytic function but also regulates expression of POR genes) makes this a critical control point in the biosynthetic pathway. The final step in chlorophyll synthesis is addition of the 18-carbon phytol side-chain to chlorophyllide. Chlorophyll b is formed from chlorophyll a by enzyme-mediated oxidation of a side-group.

The *flu* mutant of Arabidopsis is highly photosensitive and is bleached in the light as a consequence of the buildup of reactive oxygen species[58]

Plants are absolutely dependent on light to make a living. They interact with light through their pigments. For a plant, therefore, light is a Good Thing. But that's not the whole story. Plants support life on Earth by collecting the solar energy that reaches the planet's surface. Earth's atmosphere filters most of the dangerous ionizing ultraviolet radiation that would otherwise be hazardous to living matter. Nevertheless, as anyone who's been sunburnt knows, high-energy photons get through in enough intensity to make it a necessity for living organisms (including people) to equip themselves with antioxidant defences, repair mechanisms and sunblockers. The harmful effect of light is referred to as photodynamic stress or photosensitisation. Even the lower-energy visible wavelengths can be dangerous through the mediation of pigments, and this is a particular hazard for plants, for whom light capture is a central fact of life. Unless put to physiological use, light energy absorbed by chlorophyll and other tetrapyrroles can be transferred to other molecules, leading to the formation of reactive oxygen species (ROS), which in turn can result in pathological photodynamic injury to proteins and lipid membranes. The functional machinery of photosynthesis is normally organised in the membranes of the chloroplast to avoid and ameliorate pigment over-excitation. But tetrapyrrole synthesis is particularly vulnerable because so many of the later intermediates in the magnesium branch, beginning with red-brown protoporphyrin IX[59], are pigmented and therefore potential sources of photostress. For this reason, chlorophyll biosynthesis is normally under stringent control to ensure intermediates are transient[60], and is usually tightly coupled to the synthesis of carotenoids which (as discussed presently) have a crucial protective function in photostress[61]. Certain genetic mutations or chemical treatments, however, can cause the buildup of photoreactive tetrapyrrole metabolites, rendering plants highly sensitive to light damage. Some types of bleaching herbicide - for example the flurazons, which block carotenoid synthesis - work by interfering with the chlorophyll synthesis pathway and its light stress defences, thereby promoting the damaging photodynamic influence of tetrapyrrole intermediates[62]. ALA fed to plant tissue in the dark can be converted to tetrapyrrole products that accumulate to such a degree that photodynamic injury kills cells within minutes of exposure to light[63].

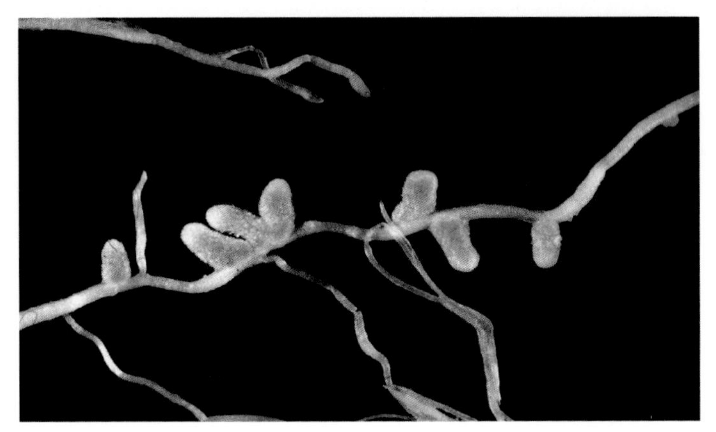

Nitrogen-fixing root nodules of the legume *Medicago*

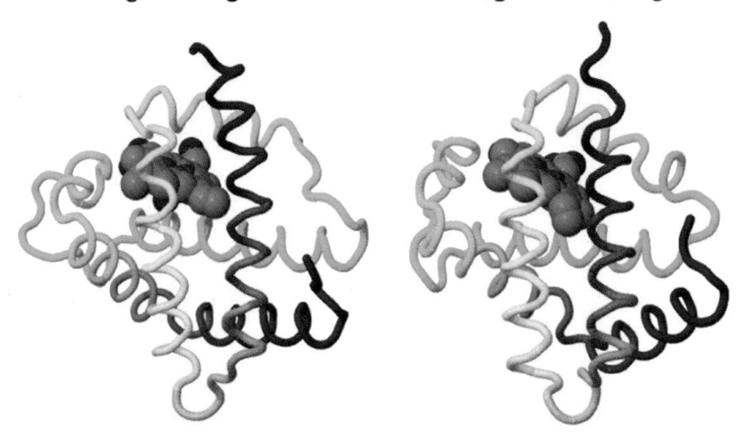

Leghaemoglobin

Bovine haemoglobin α

Plant and animal haemoglobins[64]

After the protoporphyrin IX branchpoint, insertion of iron (Fe) is the first step in the haem pathway of tetrapyrrole synthesis. Everyone's familiar with haem in the form of the haemoglobin that gives blood its red colour; but there's more to haem than meets the eye. It's a universal cofactor in living organisms, including plants, and is associated with a diversity of proteins that generally function in oxidation and reduction reactions. In a bizarre example of convergent evolution, a form of haemoglobin occurs in the nitrogen-fixing root nodules of legume plants. Leghaemoglobin in nodules regulates the availability of oxygen, which is an inhibitor of the nitrogenase enzyme that converts atmospheric nitrogen into ammonia[65]. The molecular structures of blood haemoglobin and leghaemoglobin are extraordinarily similar. Among the haem-proteins, the cytochromes are of special significance. We already met some of them as components of the Cytbf complex of the photosynthetic electron transport chain. Other cytochromes are built in to photosystem reaction centres, and they are also essential for energy generation by respiratory oxidation. Of all cytochromes, the P450 types are particularly notable (P450 refers to their original experimental detection by spectrometry based on light absorption at 450 nm). The plant cytochrome P450 superfamily comprises 200-500 diverse genes, accounting for more than 1% of all functional genes and representing the most abundant class of enzyme-encoding genes in the genome[66]. The number and variety of cytochrome P450 genes reflects the prodigious capacity of plants to make and break a great range of biomolecules (including pigments), which in turn is an essential requirement for survival in a sedentary organism. It seems clear that the multiplicity of P450s is the evolutionary outcome of diverse genetic mechanisms that favour the emergence of novel metabolic functions[67]. Among many other variants of haem as a cofactor in plant metabolism we might mention sirohaem, which plays a central role in the assimilation of nitrogen and sulphur[68].

Elevated ALA and PBG $\xrightarrow[\text{72 hours}]{\text{Sunlight + } O_2}$ Red-brown porphyrin polymers of PBG

Urine sample from a patient with acute congenital porphyria[69]

The chlorophyll-haem branchpoint in tetrapyrrole biosynthesis is a site of sensitivity in so-called retrograde signalling, the cross-talk between genes in the nucleus and chloroplasts that ensures coordinated differentiation and functioning of the photosynthetic cell. Perturbation here generates a distress signal, part of a suite of intracellular regulatory interactions that include sensitivity to reactive oxygen, levels of intermediates in chlorophyll synthesis and the state of plastid gene expression[70]. Plant responses to photosensitisation by excessive accumulation of intermediates in tetrapyrrole biosynthesis have their counterparts in animals[71]. Porphyrias are human genetic diseases caused by deficiencies of specific enzymes in the haem biosynthetic pathway. Symptoms may include photosensitivity - sufferers need to avoid bright light as otherwise they may develop painful skin lesions[72]. It has been suggested (though the idea isn't much supported these days) that porphyria is the origin of the vampire legend. The cell-killing properties of natural and chemically-synthesised tetrapyrroles and their biosynthetic precursors have been exploited in the treatment of diseases, especially cancerous tumours. Photodynamic therapy involves delivering a photosensitizing drug to the affected tissue and exposing the target to high-intensity laser light to induce localized production of reactive oxygen species which induce cell death. ALA is widely used as a photosensitizer, as are derivatives of bacteriochlorophyll.

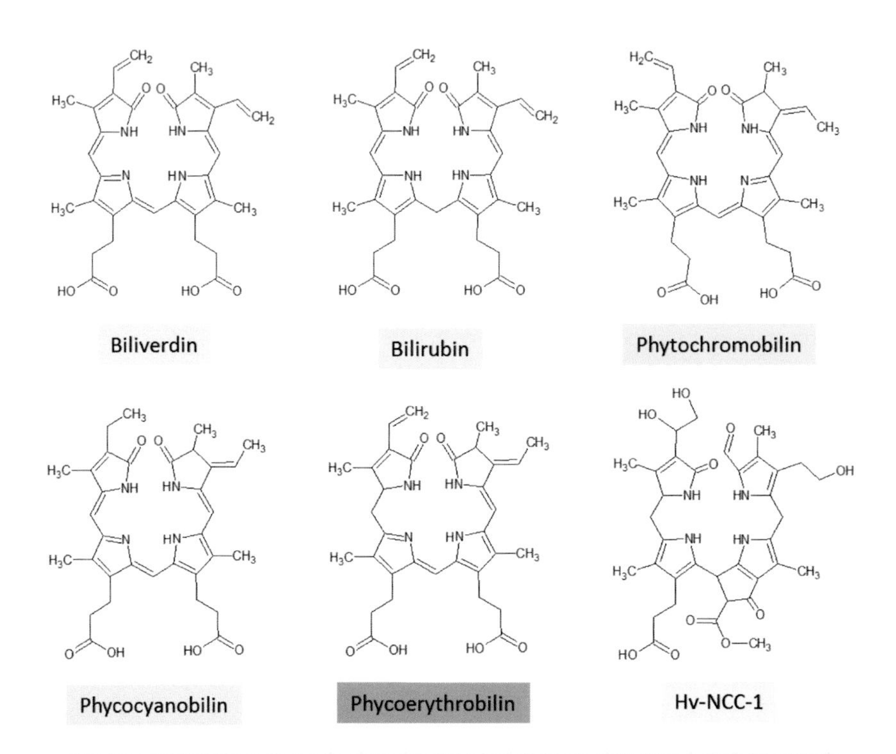

Bilins, derived from haem, except for Hv-NCC-1, which is the terminal breakdown product of chlorophyll in senescent barley leaves

Haem is the source of a number of important plant pigments. Onward metabolism of haem begins with haem oxygenase, an enzyme found in organisms across all taxonomic groupings. It opens the macrocycle of haem, releasing Fe and carbon monoxide. Haem oxygenase has long been a subject of great biomedical interest, since it has a central function in cell homeostasis and stress sensing. This makes it a prime target for the discovery of new drugs to treat conditions such as cancer and inflammation arising from pathological effects of reactive oxygen[73]. The linear tetrapyrroles derived from opening up the haem macrocycle are collectively referred to as bilins[74], taking the name from bile, into which the products of haem oxygenase are secreted. The immediate product of the enzyme, biliverdin, is the precursor of phytochromobilin, the light-sensitive cofactor of the plant photoreceptor phytochrome. Plant development is controlled by light, and phytochrome is the major sensor that detects and transduces daylength, season and light quality signals from the environment[75]. Phytochrome exists in two photoreversible forms: PR which absorbs red light (around 660 nm); and PFR which absorbs far-red (about 730 nm). The essence of phytochrome's function is red/far-red reversibility: when the phytochromobilin of PR absorbs red light it flips into the molecular form of PFR, which reverts to PR when exposed to far-red light. Far-red/red-induced alteration in phytochromobilin's molecular structure in turn leads to conformational changes in the phytochrome protein to which it is attached and initiates a cascade of physiological events resulting in photomorphogenetic responses such as the induction of flowering, the onset of dormancy or shade avoidance. Other plant bilins derived from opening the haem macrocycle include phycocyanobilin and phycoerythrobilin (phycobilins), photosynthetic pigments found alongside chlorophyll a in red algae and in cyanobacteria. By capturing light in the middle of the visible spectrum, where absorption by chlorophyll is low, they make more effective use of the attenuated light energy penetrating the water column[76]. Phycobilins are bound to specific proteins. These phycobiliproteins are organized into complex light-harvesting structures called phycobilisomes and account for the colours of red algae and cyanobacteria.

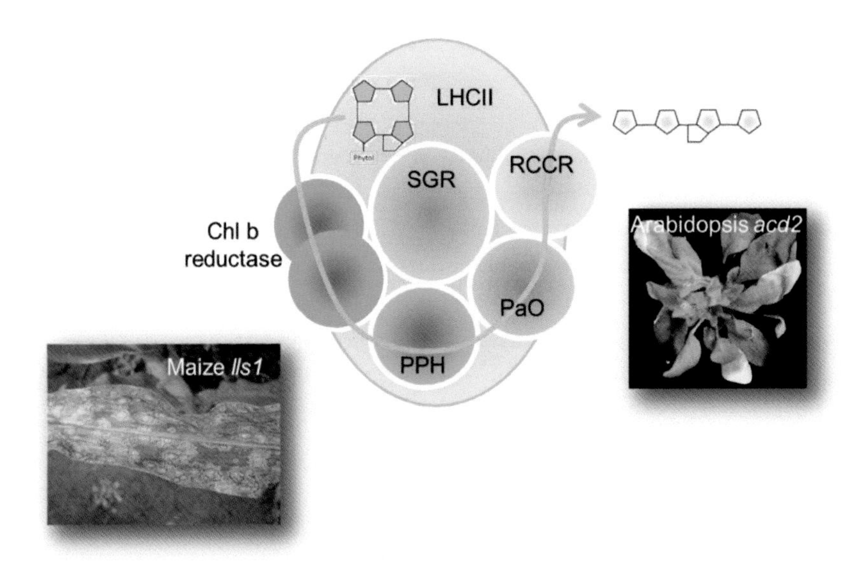

Chlorophyll is broken down into a non-photodynamic bilin by the PAO/phyllobilin pathway, catalysed by a multi-protein molecular machine. The maize cell death mutant *lls1* has a defect in the enzyme PaO, and the Arabidopsis mutant *acd2* lacks RCCR activity[77]

Bilin-like tetrapyrroles are formed during foliar senescence, a process that occurs on a global scale, when the loss of chlorophyll is the visible symptom of a change in chloroplast structure and function[78]. The capacity of the chloroplast to carry out photosynthesis declines and at the same time it becomes a source of proteins and other materials for salvage and recycling to new, growing tissues. It seems that plants deal with unwanted chlorophyll by treating it as a phototoxic compound, converting it to bilin products that are rendered harmless and finally dumped in the vacuole, the cell's all-purpose isolation unit. In this way cell integrity is preserved throughout senescence, allowing protein nitrogen to be recovered efficiently[79]. The early intermediates of chlorophyll catabolism are phaeo pigments, olive-green tetrapyrroles from which the central magnesium has been removed. Phaeo compounds are photoactive and have been used in photodynamic therapies to cause lipid peroxidation and arrest of growth in tumours. Generally, phaeo pigments occur only transiently in viable plant tissue, although a phaeophytin molecule has been identified as a primary electron acceptor in the PSII reaction centre. During chlorophyll breakdown, phaeo derivatives are converted into terminal bilins by a kind of molecular machine that keeps the photodynamic properties of intermediates in check. If the process is disrupted – as, for example, in mutants with defective chlorophyll breakdown enzymes – tissue can be photosensitised, and cells will bleach and be killed on exposure to light.

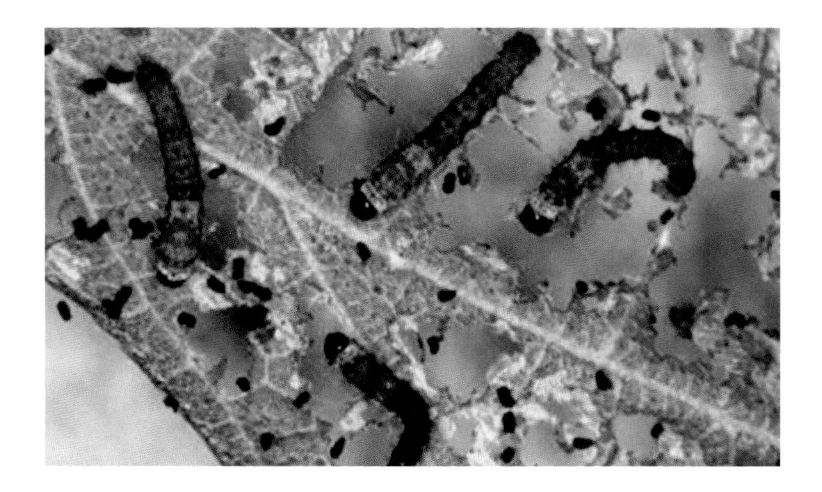

Mulberry leaf chlorophyll goes straight through a silkworm[80]

During haem turnover in animals, biliverdin, the product of haem oxygenase is converted to bilirubin which is expelled into the gut and accounts for the red-brown colour of faeces (bilins derived from blood are also responsible for the garish colours of bruises). While we're on the subject, the digestive systems of herbivores and omnivores will also contain chlorophyll from the plants that went in the front end. What happens to this pigment on its way through the gut? Normally it passes through pretty harmlessly. The typical olive-green colour of a fresh cowpat suggests there's a lot of phaeo material there. In some circumstances, however, chlorophyll derivatives penetrate the gut barrier into the bloodstream and in rare cases this can turn serious. Dietary chlorophyll getting to the skin can be bad for albino animals, because the light absorbed by the pigment will cause unpleasant dermal lesions[81]. On a global scale the fate of an awful lot of chlorophyll is to pass through an animal. In some ecosystems as much as 70% of plant biomass is disposed of this way, with interesting ecological and economic implications. For example the frass (faeces) of silkworms fed in the traditional fashion on mulberry leaves are a concentrated source of chlorophyll[82]. Until intensive silk production moved over to artificial diets, chlorophyll was a valuable by-product of the industry, supplying much of the world's needs for chlorophyll as a food colour, breath freshener and surgical adhesive. A large proportion of global chlorophyll is found in the oceans. Many planktonic animals are transparent and, when they are stuffed with chlorophylls from the algae they consume, they have to adopt special measures to avoid photodamage. In some species the wall of the gut is dark-coloured, making it opaque. Others avoid the light by undertaking a daily vertical migration in the water-column[83]. Throughout geological time a rain of chlorophyll has been constantly falling to the ocean bed, and much of it has been through the guts of animals in the food chain. This abyssal chlorophyll is thought to be the origin of the distinctive tetrapyrroles found in sediments and fossil fuels[84]. Bilin products of chlorophyll ingested by certain planktonic and krill species are substrates for luciferases, the enzymes responsible for bioluminescence in marine environments[85]. Thus the superficially unsavoury subject of guts and faeces turns out to lead in many unexpected directions, including textiles, global energy sources and the beauties of photobiology.

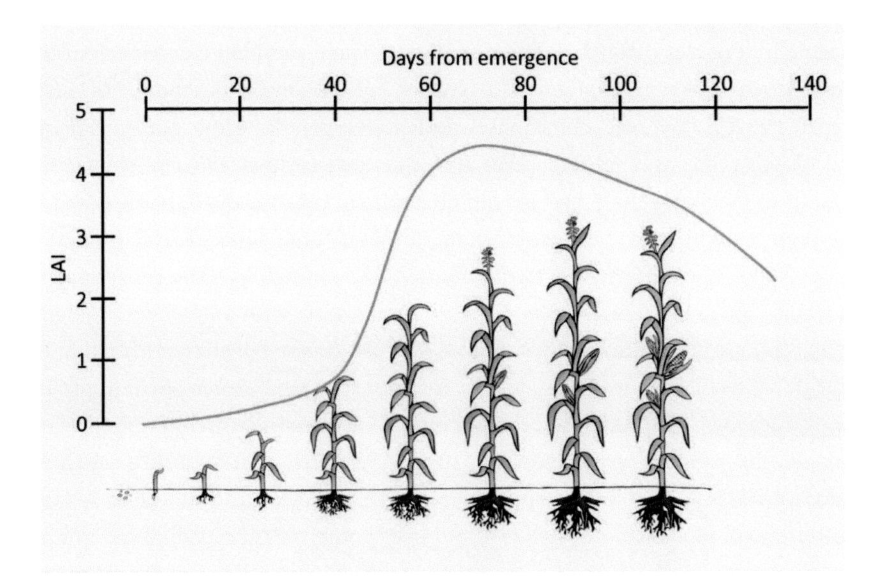

Agriculture seeks to maximise the collection, retention and utilisation of solar energy. The LAI curve (the example is maize) is a measure of canopy light capture from planting to harvest[86]

The crops that feed the world share a life cycle. First the crop must collect photons. It does this by establishing and maintaining foliage, which acts as an array of green solar panels. Then comes the transition from vegetative to reproductive growth and the formation of seeds, fruits or other harvested foodstuff. Chlorophyll is the portal through which photon energy enters the biosphere. Broadly speaking, when it comes to biological productivity, Green is Good. Intensive selection has been steadily increasing both yield and the duration of greenness in a range of agricultural species since the early decades of the twentieth century[87]. Plant breeders and agronomists, being practical and down to earth people, often use the plain and simple term 'stay-green' to describe crop varieties that remain lush and productive under the environments and management regimes of developed-world agriculture. By the end of the 1970s stay-green was becoming established explicitly as a superior characteristic and marketing feature of commercially bred grain crops, particularly maize[88]. The capture of light energy is reflected in the ebb and flow of greenness as a crop establishes, grows, ripens and yields. At the same time, the relationship of the canopy to the import and export of carbon, nutrients and water changes with stage of development. Agronomists and plant breeders have long been accustomed to judge the performance of potential and established crop varieties by eyeballing. Fortunately, the colour sensitivity of the human eye is maximal around the peak of light reflected from chlorophyll; actually, this is probably no coincidence – there is good reason to believe that evolution has spectrally tuned vision in humans and other trichromatic old-world primates[89], to the various colours of leaves. The eyes of field scientists are being supplemented by electronic gadgets, some of them mounted on drones, aircraft or satellites, that monitor vegetation colour by spectrometry and imaging[90]. Similar approaches are used for studies of natural ecosystems too.

Sorghum without (left) and with (right) the stay-green locus *Stg2* in a similar genetic background, grown under the same field conditions and exposed to terminal water deficit[91]

The story of the cereal crop sorghum in Indian agriculture represents a case study on the central significance of chlorophyll for crop productivity. Sorghum, one of the five most globally important cereals, is a staple dry land subsistence food crop for millions of resource-limited farmers in the fragile drought-prone semi-arid tropic regions of Asia, Africa and Latin America. The Indian experience over the last half-century or so shows that a combination of irrigation and policy support is effective in increasing the production efficiency of sorghum, a 'poor man's crop' that the Green Revolution in rice and wheat is often supposed to have passed by[92]. These advances are complemented by varieties with enhanced stable greenness, bred to respond to improved crop management and to tolerate the stresses of semi-arid environments[93]. Varieties that retain green leaf area have a superior capacity to maintain plant water status and grain fill, to exhibit high food and feed quality, and to resist biotic and abiotic stresses. The availability of a diversity of germplasm with superior greenness, and molecular analysis of stay-green genes, open the way to more efficient high-precision breeding strategies[94] and even, ultimately, transgenic technologies. Addressing the biological limitations to sorghum productivity is within sight; ensuring compatibility between the benefits of improved varieties and the economic, social and cultural circumstances of the people dependent on sorghum is an altogether trickier problem – which brings us to the vexed subject of the politics of chlorophyll.

The green dilemma

Chlorophyll stands at the centre of some of the great global challenges of the age. Take the example of sorghum. It's difficult to dispute the view that increasing sorghum productivity - by keeping the crop greener, for example - helps to improve the circumstances of the most disadvantaged people in some of the most difficult environments, and will contribute towards achieving United Nations Development Goals[95]. But the strategy is burdened with socio-political complexities, centred on the need for a Doubly Green Revolution, to use the title of the influential book by Conway[96], and concerns about protecting the rights, knowledge and germplasm of indigenous peoples against biopiracy and agribusiness exploitation[97]. Across the world we find the ideals of Green Revolution and Green Environmentalism in collision, a strange state of affairs when the two movements might be expected to be on the same side.

The history of the influx of physicists to the field of genetics in the middle of the last century and their influence on the birth of the molecular biology revolution is well documented[98]. Rather less familiar is the story of the decades following the 1939-45 World War in which a generation of scientists (physicists, mathematicians and engineers, as well as biologists) chose to devote their skills to agriculture (these days they go into finance). A powerful influence was the exposure of ordinary people in the developed world, for the first time, to images of mass starvation on the television screen in the corner of the room[99]. Paul Ehrlich's apocalyptic best-seller *The Population Bomb*[100] caught, and intensified, the mood of the age. Radical political movements in the West at that time also contributed to the motivations of the new agriculturalists. Above all, most of those who took this path thought they were doing something honourable and philanthropic. Surely devoting one's professional life to defeating famine was the right thing to do.

From the perspective of the next century, this attitude looks at best naive and, when it comes up against the anti-technological politics of contemporary popular Green movements, positively shameful. Many of the original idealistic feed-the-world scientists are still in business but having to contend with sometimes bewildering and frustrating hostility from groups that ought to share their ideals. Among the serious consequences are stagnation of financial and societal support for open public-good crop research, driving more and more of this work into the private corporate sector. It is difficult to see how this will benefit the poor and disadvantaged[101]. These developments are all the more vexing now that at last we are acquiring the tools to do the job. That these tools are technological (particularly biotechnological) is, to a large measure, the problem. To get them accepted by those who should benefit from agricultural progress needs trust. Lack of nuance and sensitivity on the part of agribusiness and governments, and a general failure of communication by a scientific community driven by the best intentions, can land the average green-sympathetic jobbing crop scientist with a crisis of conscience. Is she doing good or harm?

NOTES AND SOURCES

[27] Public domain (CC0) illustration by Lilla Frerichs (www.publicdomainpictures.net/en/view-image.php?image=56748&picture=tropical-leaf-fan [accessed 25 May 2018]).
[28] Image from *The Booke of the Rosary of Philosophers* (16th Century; Glasgow University Library MS. Ferguson 210).
[29] Dobbs BJT. 1975. *The Foundations of Newton's Alchemy or "The hunting of the greene lyon".* Cambridge: University Press.
[30] Newman WR. 2010. Newton's early optical theory and its debt to chymistry. In: Jacquart D, Hochmann M (eds) *Lumière et Vision dans les Sciences et dans les Arts. De l'Antiquité au XVIIe siècle.* Droz, Genf, pp.283-307.
[31] Hangarter RP, Gest H. 2004. Pictorial demonstrations of photosynthesis. Photosynthesis Research 80: 421-425.
[32] Stock photo 42569223, licensed by www.123rf.com.
[33] Goldsworthy A. 1987. Why did Nature select green plants? Nature 328: 207—208.
[34] DasSarma S. 2006. Extreme halophiles are models for astrobiology. Microbe-American Society for Microbiology 1: 120—126.
[35] Chlorins and porphyrins have 24 atoms in the macrocycle, corrinoids 23. Tetrapyrroles readily interconvert spontaneously, and have been the subjects of a diversity of chemical modifications, many of which have industrial and biomedical uses (Senge MO, Sergeeva NN. 2006. Metamorphosis of tetrapyrrole macrocycles. Angewandte Chemie International Edition 45: 7492-7495).
[36] Woodward RB. 1961. The total synthesis of chlorophyll. Pure and Applied Chemistry 2: 383-404.
[37] Woodward RB. 1973. The total synthesis of vitamin B_{12}. Pure and Applied Chemistry 33: 145-178.
[38] Chlorophylls c and d, of algal origin, are further variants of the chlorophyll a structure: c has combinations of $-CH_3$, $-C_2H_5$ and $-COOCH_3$ groups on ring II; d has –CHO instead of $-CHCH_2$ on ring I.
[39] Taniguchi M, Ptaszek M, Chandrashaker V, Lindsey JS. 2017. The porphobilinogen conundrum in prebiotic routes to tetrapyrrole macrocycles. Origins of Life and Evolution of Biosphere 47: 93-119.
[40] Taniguchi et al. (2017).
[41] Sessions AL, Doughty DM, Welander PV, Summons RE, Newman DK. 2009. The continuing puzzle of the great oxidation event. Current Biology 19: R567–R574.
[42] Fujita Y, Tsujimoto R, Aoki R. 2015. Evolutionary aspects and regulation of tetrapyrrole biosynthesis in cyanobacteria under aerobic and anaerobic environments. Life 5: 1172-1203.
[43] Synthetic tetrapyrrole-like chemicals have a wide range of technological and clinical applications, for example as dyes and as photosensitisers for medical treatments. About a quarter of all industrially produced organic pigments are derivatives of the porphyrin-related phthalocyanines (Walter MG, Rudine AB, Wamser CC. 2010. Porphyrins and phthalocyanines in solar photovoltaic cells. Journal of Porphyrins and Phthalocyanines 14: 759-792; Huang H, Song W, Rieffel J, Lovell JF. 2015. Emerging applications of porphyrins in photomedicine. Frontiers in Physics 3: 23).

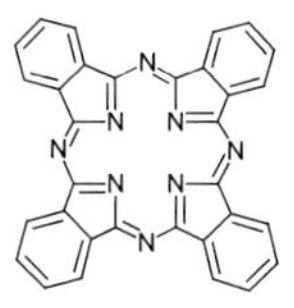

Phthalocyanine

[44] Granick S. 1951. Biosynthesis of chlorophyll and related pigments. Annual Review of Plant Physiology 2: 115--144.
[45] Images from various public domain sources.
[46] Jones R, Ougham H, Thomas H, Waaland S. 2013. *Molecular Life of Plants*. Chichester: Wiley-ASPB.
[47] Electron micrograph from a study by Helen Ougham and the author. Plastoglobules (lipid droplets) greatly increase in size, number and pigmentation when chloroplasts differentiate into chromoplasts (the plastids of fruits and flowers) and gerontoplasts (the plastids of senescing leaves).
[48] Haeckel is best known for his theory of recapitulation (ontogeny repeats phylogeny) and his beautifully illustrated publications on invertebrates. He was a controversial figure and his reputation remains a subject of debate (Richards RJ. 2009. Haeckel's embryos: fraud not proven. Biology and Philosophy 24: 147–154). Schimper was the first to propose the endosymbiotic origin of chloroplasts (Schimper AFW. 1883. Über die Entwicklung der Chlorophyllkörner und Farbkörper. Botanische Zeitung 41: 105–162).
[49] Wobbe L, Bassi R, Kruse O. 2016. Multi-level light capture control in plants and green algae. Trends in Plant Science 31: 55--68.
[50] Murchie EH, Lawson T. 2013. Chlorophyll fluorescence analysis: a guide to good practice and understanding some new applications. Journal of Experimental Botany 64: 3983-3998.
[51] Kashiyama Y, Tamiaki H. 2014. Risk management by organisms of the phototoxicity of chlorophylls. Chemistry Letters 43: 148--156.
[52] *R. viridis* reaction centre crystal structure from RCSB PDB (code 2prc), rendered in FirstGlance in Jmol (https://bioinformatics.org/firstglance/fgij/ [accessed 20 June 2018]). Comparison of polypeptides of bacterial, PSI and PSII reaction centres from Jones et al. (2012), reproduced with permission from the publisher, Wiley.
[53] Deisenhofer J, Michel H. 1989. The photosynthetic reaction center from the purple bacterium *Rhodopseudomonas viridis*. Science 245: 1463-1473.
[54] Subunit of *Pisum sativum* PSII light-harvesting antenna from RCSB PDB (code 2bhw), rendered in PDBTM (http://pdbtm.enzim.hu/ [accessed 9 August 2018]). Three helices of the light-harvesting protein spanning the membrane are associated with 6 chlorophyll a, 6 chlorophyll b and a pair of lutein molecules.
[55] Brzezowski P, Richter AS, Grimm B. 2015. Regulation and function of tetrapyrrole biosynthesis in plants and algae. Biochimica et Biophysica Acta Bioenergetics 1847: 968-985.
[56] Granick S. 1948. Protoporphyrin 9 as a precursor of chlorophyll. Journal of Biological Chemistry 172: 717-727.
[57] Armstrong GA. 1998. Greening in the dark: light-independent chlorophyll biosynthesis from anoxygenic photosynthetic bacteria to gymnosperms. Journal of Photochemistry and Photobiology B: Biology 43: 87-100.
[58] Laloi C, Havaux M. 2015. Key players of singlet oxygen-induced cell death in plants. Frontiers in Plant Science 6: 39.

[59] Granick (1948).
[60] Brzezowski et al. (2015).
[61] Park JH, Tran LH, Jung S. 2017. Perturbations in the photosynthetic pigment status result in photooxidation-induced crosstalk between carotenoid and porphyrin biosynthetic pathways. Frontiers in Plant Science 8: 1992.
[62] Kashiyama and Tamiaki (2014).
[63] Rebeiz CA, Montazer-Zouhoor A, Jopen HJ, Wu SM. 1984. Photodynamic herbicides: concepts and phenomenology. Enzyme Microbiology and Technology 6: 390-401.
[64] Root nodule illustration from https://commons.wikimedia.org/wiki/File: Medicago_italica_root_nodules_2.JPG ([accessed 19 June 2018]; Creative Commons licence CC BY-SA). Leghaemoglobin and bovine haemoglobin structures (PDPs 1bin, 2qss respectively) rendered in FirstGlance in Jmol.
[65] Becana M, Klucas RV. 1992. Oxidation and reduction of leghemoglobin in root nodules of leguminous plants. Plant Physiology 98: 1217-1221.
[66] Jun XU, Wang XY, Guo WZ. 2015. The cytochrome P450 superfamily: key players in plant development and defense. Journal of Integrative Agriculture 14: 1673-1686.
[67] Liu Z, Tavares R, Forsythe ES, André F, Lugan R, Jonasson G, Boutet-Mercey S, Tohge T, Beilstein MA, Werck-Reichhart D, Renault H. 2016. Evolutionary interplay between sister cytochrome P450 genes shapes plasticity in plant metabolism. Nature Communications 7: 13026.
[68] Tripathy BC, Sherameti I, Oelmüller R. 2010. Siroheme: an essential component for life on earth. Plant Signaling and Behavior 5: 14-20.

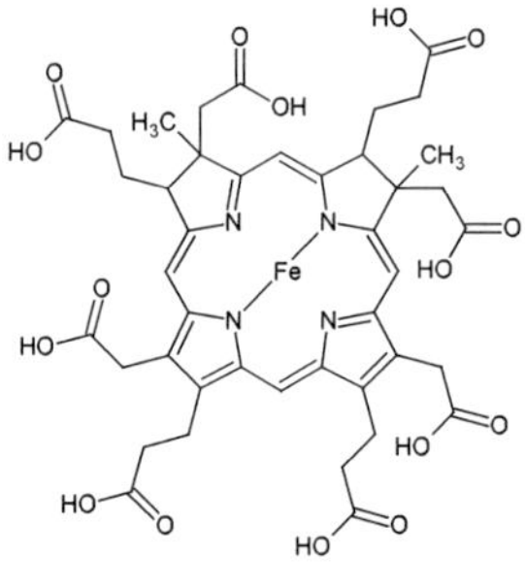

Sirohaem

[69] Illustration under Creative Commons CC BY, from https: //commons.wikimedia.org/wiki/File:Urine_of_patient_with_porphyria.png [accessed 20 June 2018].
[70] Chan KX, Phua SY, Crisp P, McQuinn R, Pogson BJ. 2016. Learning the languages of the chloroplast: retrograde signaling and beyond. Annual Review of Plant Biology 67: 25-53.
[71] Dayan FE, Dayan EA. 2011. Porphyrins: One Ring in the Colors of Life: A class of pigment molecules binds King George III, vampires and herbicides. American Scientist 99: 236-243.
[72] Thadani H, Deacon A, Peters T. 2000. Diagnosis and management of porphyria. British Medical Journal 320: 1647–1651.
[73] Motterlini R, Foresti R. 2014. Heme oxygenase-1 as a target for drug discovery. Antioxidants and Redox Signaling 20: 1810-1826.
[74] Plant bilins are referred to as phyllobilins.

[75] Kami C, Lorrain S, Hornitschek P, Fankhauser C. 2010. Light-regulated plant growth and development. Current Topics in Developmental Biology 91: 29–66.
[76] Croce R, Van Amerongen H. 2014. Natural strategies for photosynthetic light harvesting. Nature Chemical Biology 10: 492.
[77] For an overview of chlorophyll breakdown, see Kuai et al. (2017). The SGR protein is necessary for stabilising the enzymes of chlorophyll degradation and binding to pigment-protein complexes in the thylakoid. Mutants lacking SGR remain green during senescence. SGR also removes Mg from the chlorophyll macrocycle to form phaeo pigments (Shimoda et al. 2016). Chlorophyll b is degraded after first being converted to chlorophyll a by Chl b reductase. The phytol side-chain is removed from phaeophytin a by PPH and the macrocycle of the product, phaeophorbide a, is cleaved between pyrrole rings I and II by PaO. The resulting bilin is a red catabolite (RCC) and is reduced to a colourless, fluorescent product (FCC) by RCC reductase. Phaeophorbide and RCC are highly phototoxic and blocking their metabolism (as in the lesion-mimic mutants *lls* and *acd2*; Bruggeman et al. 2015) results in photosensitive cell death (Kuai B, Chen J, Hörtensteiner S. 2017. The biochemistry and molecular biology of chlorophyll breakdown. Journal of Experimental Botany 69: 751–767; Shimoda, Y., Ito, H., and Tanaka, A. (2016). Arabidopsis STAY-GREEN, Mendel's green cotyledon gene, encodes magnesium-dechelatase. Plant Cell 28: 2147–2160; Bruggeman Q, Raynaud C, Benhamed M, Delarue M. 2015. To die or not to die? Lessons from lesion mimic mutants. Frontiers in Plant Science 6: 24).

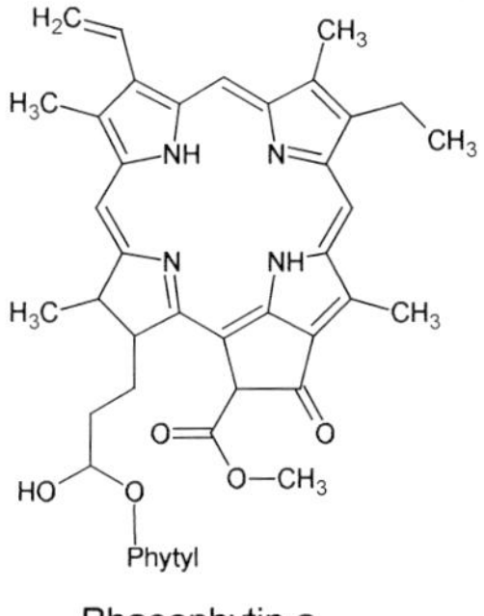

Phaeophytin a

[78] Ougham H, Hörtensteiner S, Armstead I, Donnison I, King I, Thomas H, Mur L. 2008. The control of chlorophyll catabolism and the status of yellowing as a biomarker of leaf senescence. Plant Biology 10 (supplement 1): 4–14.
[79] Diaz-Mendoza M, Velasco-Arroyo B, Santamaria ME, González-Melendi P, Martinez M, Diaz I. 2016. Plant senescence and proteolysis: two processes with one destiny. Genetics and Molecular Biology 39: 329-338.
[80] Public domain photograph from Department of Entomology, University of Nebraska-Lincoln (https://entomology.unl.edu/silkworm [accessed 20 June 2018]).
[81] For example, Tapper BA, Lohrey E, Hove EL, Allison RM. 1975. Photosensitivity from chlorophyll-derived pigments. Journal of the Science of Food and Agriculture 26: 277-284; Campbell WM, Dombroski GS, Sharma I, Partridge AC, Collettt MG. 2010. Photodynamic chlorophyll a metabolites, including phytoporphyrin (phylloerythrin), in the blood of photosensitive livestock: Overview and measurement. New Zealand Veterinary Journal 58: 146-154. There is increasing evidence that phyllobilins may also cross the gut barrier with

physiological consequences (Pérez-Gálvez A, Roca M. 2017. Phyllobilins: a new group of bioactive compounds. Studies in Natural Products Chemistry 52: 159-191).

[82] Ziran H, Pingxiong J, Guodong F, Guangxue L, Yongmin C, Fengming S, Xiangrui Z, Quan L, Guozhang P. 2016. Research progress on production of chlorophyll and other products with silkworm excrement and industrial development over the past 50 years. Animal Husbandry and Feed Science 8: 174-178.

[83] For example, Nemoto T. 1968. Chlorophyll pigments in the stomach of euphausiids. Journal of the Oceanographical Society of Japan 24: 253-260; Szymczak-Żyla M, Kowalewska G, Louda JW. 2008. The influence of microorganisms on chlorophyll a degradation in the marine environment. Limnology and Oceanography 53: 851-862.

[84] Keely BJ. 2006. Geochemistry of chlorophylls. In: Grimm B, Porra RJ, Rüdiger W, Scheer H. (eds) *Chlorophylls and Bacteriochlorophylls.* Advances in Photosynthesis and Respiration 25: 535-561. Springer, Dordrecht: Springer.

[85] Haddock SH, Moline MA, Case JF. 2010. Bioluminescence in the sea. Annual Review of Marine Science 2: 443-493. Here's the phyllobilin-like luciferin of dinoflagellates and the *Euphausia* spp. shrimps that feed on them. Note that the chlorophyll macrocycle has been broken between pyrroles IV and I.

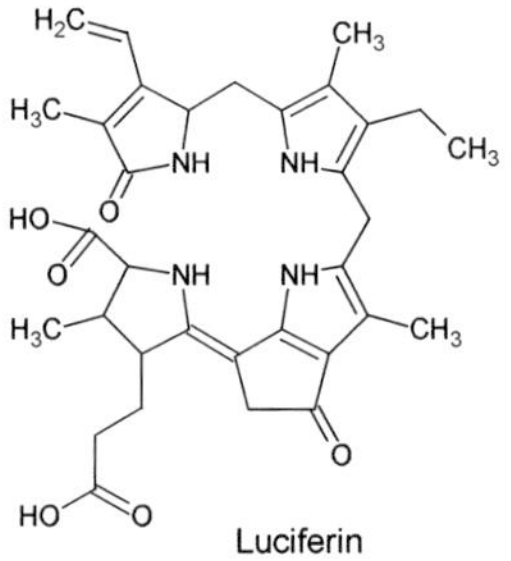

Luciferin

[86] LAI is leaf area index, the area of foliage per unit area of ground below it. Crop yield depends substantially on the speed with which the canopy closes (that is, achieves an LAI of more than 1) and on the duration of green area (Thomas H, Ougham H. 2014. The stay-green trait. Journal of Experimental Botany 65: 3889-3900). LAI curve based on data from FAO, vector illustration of maize crop stages image 40922912 under standard licence from www.123rf .com.

[87] Thomas H, Smart CM. 1993. Crops that stay green. Annals of Applied Biology 123: 193-219; Duvick DN, Smith JSC, Cooper M. 2004. Long-term selection in a commercial hybrid maize breeding program. Plant Breeding Reviews 24: 109-152.

[88] Thomas and Ougham (2014).

[89] Melin AD, Hiramatsu C, Fedigan LM, Schaffner CM, Aureli F, Kawamura S. 2012. Polymorphism and adaptation of primate colour vision. In: Pontarotti P (ed) *Evolutionary Biology: Mechanisms and Trends.* Heidelberg: Springer, pp. 225-241.

[90] For example, Rodriguez-Moreno F, Zemek F, Kren J, Pikl M, Lukas V, Novak J. 2016. Spectral monitoring of wheat canopy under uncontrolled conditions for decision making purposes. Computers and Electronics in Agriculture 125: 81-88.

[91] Many thanks to Andy Borrell, leader of research on stay-green sorghum in the University of Queensland, for providing the illustration.

[92] Janaiah A, Achoth L, Bantilan C. 2005. Has the Green Revolution bypassed coarse cereals? The Indian experience. Electronic Journal of Agricultural and Development Economics 2: 20-31.
[93] Jordan DR, Mace ES, Cruickshank AW, Hunt CH, Henzell RG. 2011. Exploring and exploiting genetic variation from unadapted sorghum germplasm in a breeding program. Crop Science 51: 1444-1457.
[94] Reddy NRR, Ragimasalawada M, Sabbavarapu MM, Nadoor S, Patil JV. 2014. Detection and validation of stay-green QTL in post-rainy sorghum involving widely adapted cultivar, M35-1 and a popular stay-green genotype B35. BMC Genomics 15: 909.
[95] http://www.un.org/millenniumgoals/; http://www.un.org/sustainabledevelopment/sustainable-development-goals/ [accessed 22 June 2018].
[96] Conway G. 1999. *The Doubly Green Revolution*. Cornell University Press.
[97] Thompson CB. 2012. Alliance for a Green Revolution in Africa (AGRA): advancing the theft of African genetic wealth, Review of African Political Economy 39: 345-350.
[98] Keller EF. 1990. Physics and the emergence of molecular biology: A history of cognitive and political synergy. Journal of the History of Biology 23: 389-409.
[99] Franks S. 2014. *Reporting Disasters: Famine, Aid, Politics and the Media*. Hurst.
[100] Ehrlich P. 1968. *The Population Bomb.* Ballantine: Sierra Club.
[101] Spielman DJ. 2007. Pro-poor agricultural biotechnology: Can the international research system deliver the goods? Food Policy 32: 189–204; Asafu-Adjaye J et al. 2015. *An Ecomodernist Manifesto*. http://www.ecomodernism.org/ [accessed 22 June 2018]; Lynas M. 2018. *Seeds of Science. Why We Got It So Wrong On GMOs*. London: Bloomsbury.

3. GOLD

Concerning the yellow, orange and red carotenoids[102]

That heartbreaking second when it all got together: the sugars and the acids and the ultraviolets, and the next thing you knew there were tangerines and string quartets.

Edward Albee (*Seascape*)

INTO my heart on air that kills
From yon far country blows:
What are those blue remembered hills…?[103]

AE Housman (*A Shropshire Lad*)

Frits Warmolt Went was one of the giants of twentieth century plant science[104]. He is remembered particularly for the discovery and development of the auxin family of plant hormones, and for his theory of directional plant growth in response to light and gravity. A no less significant research interest was atmospheric pollution - 'air that kills' - which led Went to propose that the blue haze which gives its name to the Blue Ridge Mountains of the Appalachians and the Blue Mountains in Australia was caused by volatiles of plant origin[105]. This intuition was extremely far-sighted; a great deal of subsequent research has shown the scale of the phenomenon. It's now known that terrestrial vegetation, the principal non-industrial source of biogenic volatile organic compounds, releases about a billion tons of carbon per annum into to the global atmosphere, equivalent to an astonishing 1–2% of the net primary production of land plants[106]. The largest fraction of these volatiles is accounted for by isoprene and monoterpenes emitted by leaves. They are thought to provide cells with protection from reactive oxygen generated in chloroplasts under stress, and to act as a herbivory alarm signal and defence response. Isoprene and monoterpenes are also potent greenhouse gases. For present purposes, they are of further significance because they are biomolecules belonging to the same broad structural family as the carotenoid pigments. The general name for this varied and multifunctional family is the terpenoids or isoprenoids (I'll stick to terpenoids). I'll focus on the colourful members of the family, but we'll meet an exotic diversity of relations too as I tell the tale of the little gold pigment.

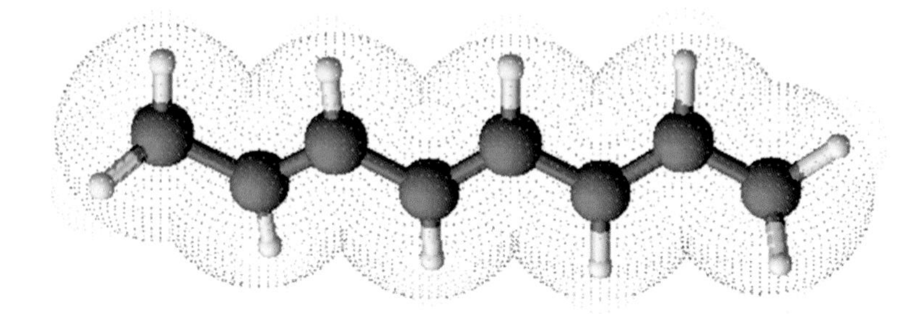

Delocalised electron cloud

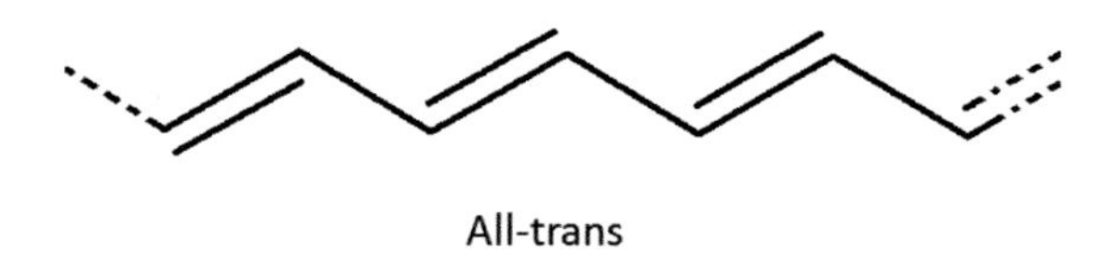

All-trans

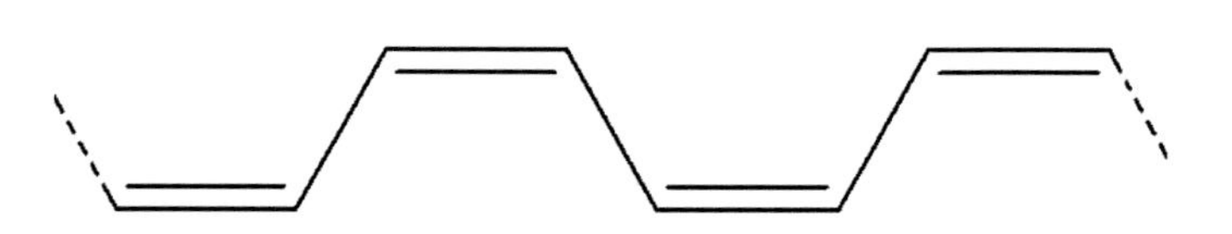

All-cis

Features of conjugated bonding systems

In 1831 the German pharmaceutical chemist Heinrich Wilhelm Ferdinand Wackenroder published his research on 'a yellow fatty oil and carotin' extracted from carrots. The subject of Wackenroder's doctoral dissertation (1826) had been the search for anthelminthics of plant origin. This led him to analyse carrot juice, which reputedly had some bioactivity. He was able eventually to obtain a few small red crystals of what we now know as β-carotene (and which has negligible antihelminthic properties). The sub-family of terpenoids to which carotene belongs is generally referred to as the carotenoids[107]. The chemical structures of major carotenoids were subsequently determined in the laboratory of Richard Willstätter, he of chlorophyll chemistry fame, between 1907 and 1913[108]. Carotenoids have characteristic conjugated (alternating double-single) bonding systems, variations in which determine the colour of the carotenoid[109]. According to quantum-mechanical theory, the wavelength of the photon absorbed by a molecule with a conjugated system of n double bonds (2n carbon atoms) is approximately proportional to n. The theory applied to terpenoids is supported by the observation that molecules where $n < 8$ are colourless to the eye, absorbing only in the UV; whereas, for each additional double bond, photons of increasing wavelength are absorbed and the colour of the molecule ranges progressively from yellow to orange to red. The conjugated bond structure of carotenoids makes them potent antioxidants. There's a theory that ageing and all manner of diseases are the result of damage by reactive forms of oxygen (ROS) and free radicals[110]. Carotenoids are essential for life, and among their vital functions in living cells is defence against the buildup of these harmful products. The Mediterranean diet, rich in tomatoes and suchlike, owes its rumoured health benefits to the presence of lots of carotenoids. Animals are unable to synthesise carotenoids and rely, directly or indirectly, on plants in the diet to meet their requirements. Vitamin A is the product of carotenoid metabolism[111]. The colours of many birds (such as flamingos), fish (salmon, for example) and invertebrates (ladybirds, lobsters) are derived from carotenoids[112].

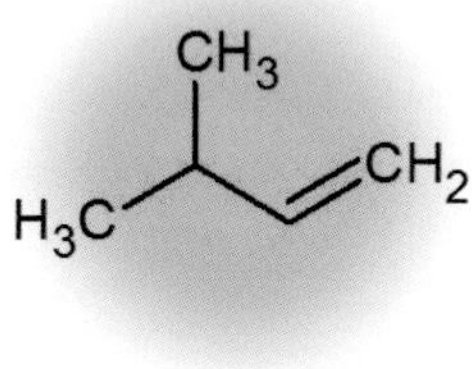

Prenyl group

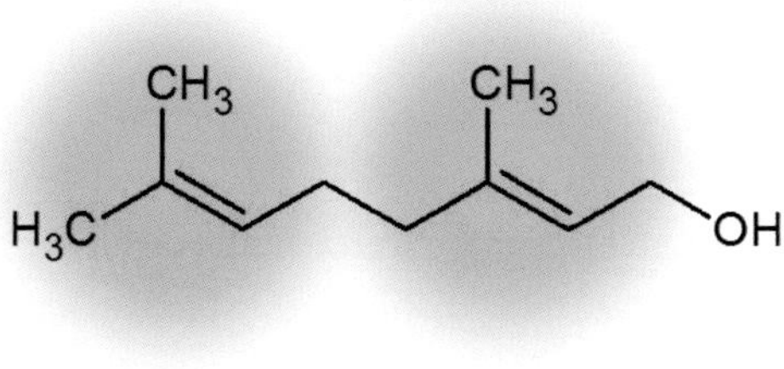

Geraniol
(a monoterpenoid)

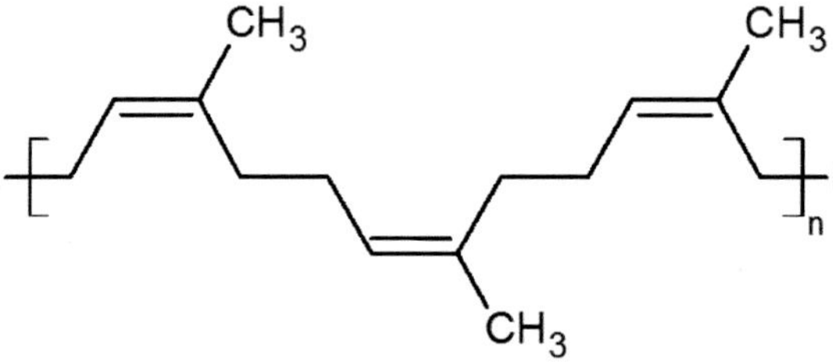

Natural rubber

Terpenoids are polymers made from prenyl units

It's estimated that there are more than 25,000 terpenoid constituents in plants, of which over 600 are carotenoids. As well as carotenoids and related compounds, the terpenoids include a structurally varied group of membrane-modifiers, toxins, antifeedants, attractants, hormones, vitamins and other secondary products. The term terpenoids is taken from terpenes, which in turn are named for the volatile constituents of turpentine, the organic solvent produced by distilling pine tree resin. The fundamental structural unit of terpenoid molecules is isoprene, which has a branched five-carbon (C5) backbone. Although, as we've seen, plants can make and emit isoprene, it is not itself an intermediate in terpenoid synthesis. Instead, the repetitive C5 module of terpenoid structure is an isomer of isoprene, the prenyl group. Terpenoids are classified by the number of multiples of the C5 unit that make up the molecule. Hemiterpenoids are C5, monoterpenoids are C10, sesquiterpenoids C15, diterpenoids C20, sesterpenoids C25, triterpenoids C30 and tetraterpenoids C40. And so on, up to macromolecule sizes; for example natural rubber is a terpenoid polymer consisting of 11,000 to 20,000 C5 units. Terpenoids that have lost one or more carbons, and so are no longer C5 multiples, are called norterpenoids. And meroterpenoids are natural products of mixed biosynthetic origins that are partially derived from terpenoids. The phytol side-chain of chlorophyll is a diterpenoid derivative. Some proteins have special membrane-associated functions as a consequence of prenylation, the addition of C15 or C20 terpenoid side-chains[113].

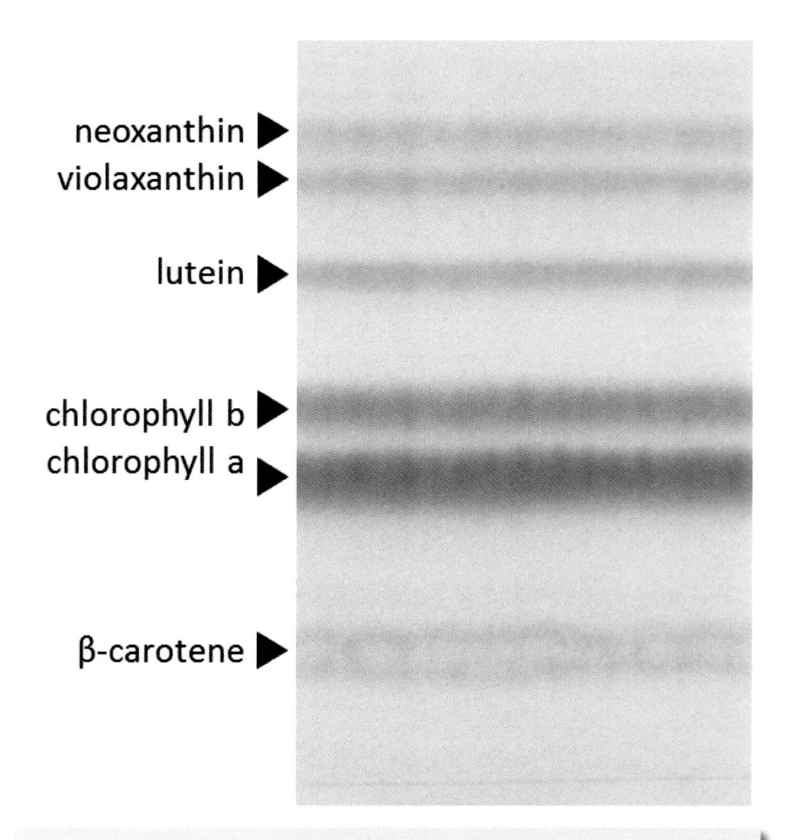

Pigments of a green leaf, separated by thin-layer chromatography[114]

Carotenoids are tetraterpenoids (C40). There are two types of carotenoid. The carotenes are constructed from carbon and hydrogen only, whereas the molecular structures of xanthophylls also include oxygen. Green photosynthetic tissues contain three major xanthophylls - neoxanthin, violaxanthin and lutein - as well as β-carotene. The carotenoid compositions of flowers and fruits can be extremely complex, with sometimes exotic variations on basic terpenoid chemical structures[115]. β-carotene is the most prominent carotenoid of almost all classes of algae. The major xanthophyll of cyanobacteria and red algae is zeaxanthin. The pigment composition of green algae resembles that of the leaves of land plants[116]. In the evolutionary history of photosynthesis, it's likely that anoxygenic photosynthetic organisms are ancestral to oxygen-producing cyanobacteria and that present-day purple bacteria are physiologically closest to the most ancient phototrophs[117]. The pigment of purple bacteria is the diterpenoid retinal which resides at the reaction centre of the light-intercepting pigment-protein complex bacteriorhodopsin[118]. We've seen in the case of tetrapyrroles that it's difficult to reconstruct the early evolution of photosynthesis. Nevertheless, it seems clear that terpenoid-derived pigments continued to carry out the function of intercepting light in the middle of the visible spectrum while chlorophyll adopted an increasingly prominent role in light capture as well as charge separation, leading to modern plants, including the land flora. Meanwhile retinal, bound to opsin proteins, went on to become the photoreceptor for animal vision. There's a feeling of some kind of deep principle at work here, uniting the eyes of the creatures that became our species with the colours of the photosynthetic organisms that surrounded and stimulated us through billions of years of evolution.

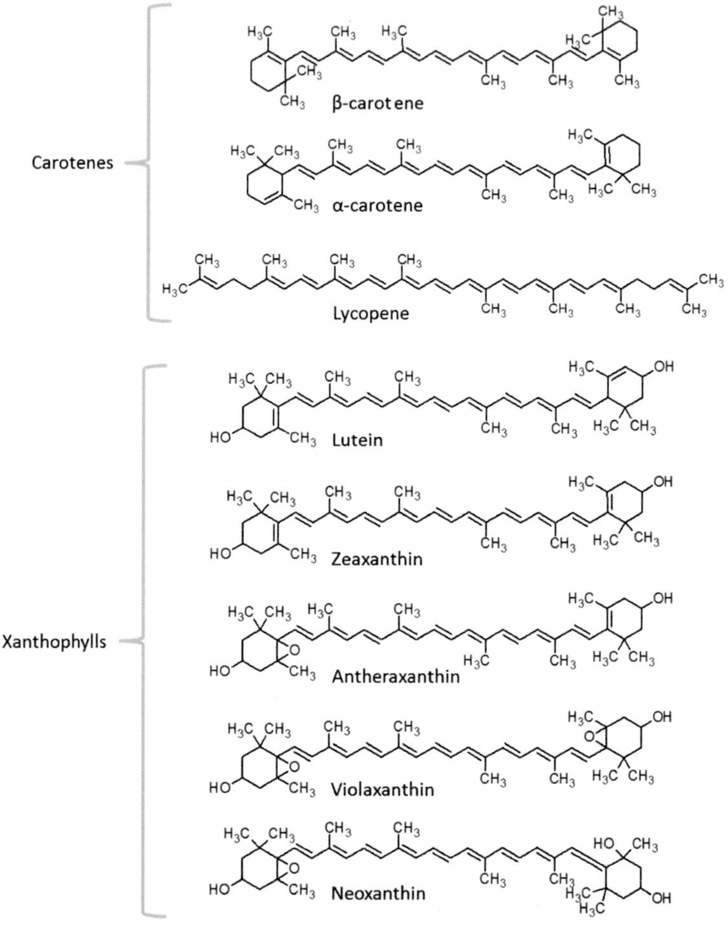

Carotenes

β-carotene

α-carotene

Lycopene

Xanthophylls

Lutein

Zeaxanthin

Antheraxanthin

Violaxanthin

Neoxanthin

Some common carotenoid pigments

Let's look in more detail at the structures of some of the commonest carotenoids. Lycopene is the pigment responsible for the bright red colour of tomatoes and capsicums and, in combination with xanthophylls and other carotenes, the orange colours of marigold flowers[119]. It has a backbone consisting of a chain of 30 carbons, linked by a conjugated bonding system, to which are attached 10 methyl (-CH_3) groups. The hydrogen atoms attached to the carbons at either end of each double bond are in the trans configuration, as is usual for most carotenoids (rubber, by contrast, is all-cis). The structure of β-carotene is basically similar to lycopene, but the ends of the C30 chain are cyclised. There are β-carotene molecules located in PSI and PSII reaction centres. They have a structural role in chlorophyll-protein complex assembly, and may make a minor contribution to light capture, but it is likely that their main job is to deal with the ROS hazard intrinsic to charge-separation[120]. The chemical structures of lycopene and β-carotene were determined in 1930 in the laboratory of the Swiss chemist Paul Karrer, who was awarded the Nobel Prize in 1937 for his work on the pigments and vitamin A[121]. One of the most abundant carotenoids is lutein, a gold-coloured xanthophyll found in all green photosynthetic tissues and in many yellow flowers. Its structure is related to that of β-carotene with the addition of a hydroxyl (-OH) group on each of the terminal rings. The light-harvesting antenna of PSII contains two lutein molecules which, like the reaction centre β-carotenes, defend against ROS and play a probable structural role. They are, however, not indispensable, since mutants lacking lutein have perfectly normal levels of chlorophylls and apparently fully functional light-harvesting structures[122]. In these mutants, lutein is replaced by violaxanthin, an intrinsic xanthophyll of green photosynthetic tissue distinguished by the epoxide groups on its terminal rings. Violaxanthin is biosynthetically related to zeaxanthin and antheraxanthin, and the three pigments are components of the xanthophyll cycle, a mechanism that regulates the amount of light energy that reaches the photosynthetic reaction centres. A notable feature of the xanthophyll neoxanthin found in the chloroplasts of all plants with chlorophyll b is a bond in the cis configuration at the 9' position adjacent to one of the terminal rings. The neoxanthin of non b-containing algae is the all-trans isomer[123].

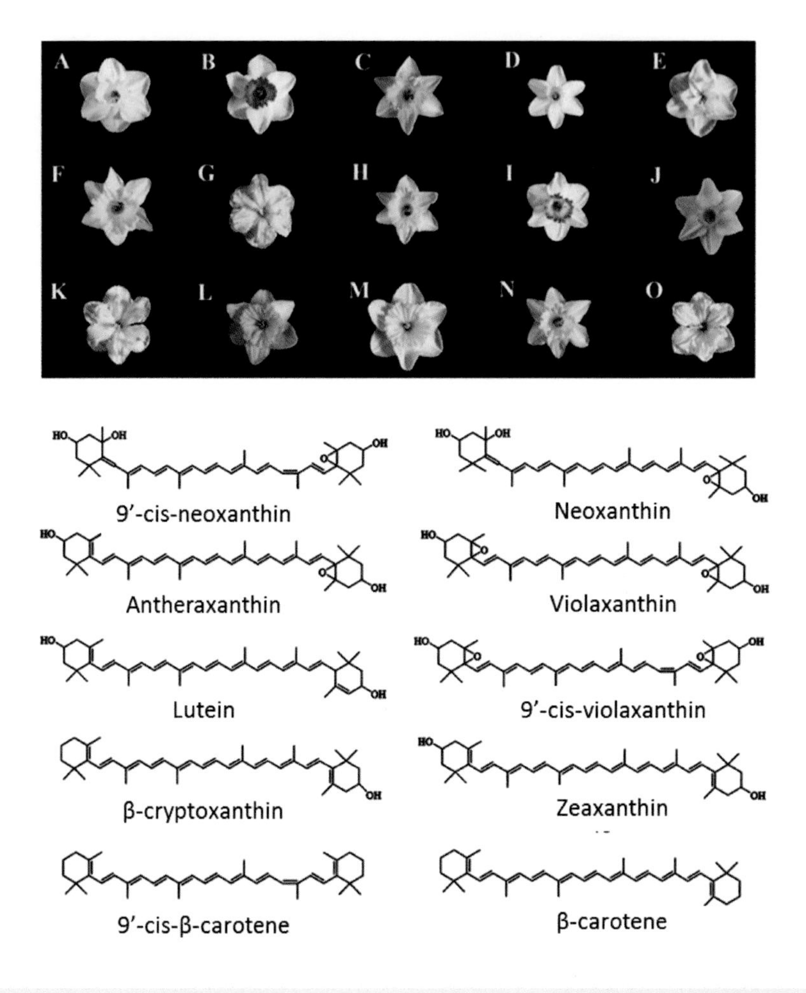

The range of *Narcissus* flower colours and their carotenoid pigments (pigments are all-trans unless otherwise indicated)[124]

I'm writing this on March 1st, St David's day, and it's satisfying to be able to use the national flower of Wales, the daffodil (*Narcissus*), as an appropriate case study of floral carotenoid diversity. Lutein is the major pigment and additionally, depending on daffodil variety and *Narcissus* species, there's a wide variety of carotenes, their isomeric forms and epoxy derivatives[125]. Looking across the range of pigment structures, not just in *Narcissus* but throughout the floral organs of angiosperms[126], one gets the impression of a core set of carotenes and xanthophylls (perhaps reflecting the common ancestry of flowers and leaves in the evolution of the shoot systems of land plants[127]) that have been tinkered with, probably under selection pressure from pollinators and, latterly, human cultivators. Elsewhere I've argued that the success of plants, seemingly fragile and unable to get out of the way, is at least in part a consequence of the capacity for this kind of speculative metabolic doodling[128]. Terpenoids are the largest group of volatiles that contribute to floral fragrances. An example is the monoterpene geraniol, a bee attractant and major constituent of rose oil. On the whole, showy, highly coloured flowers (such as a host of golden daffodils) rely on visual stimulation to interact with pollinators and tend not to be highly perfumed; whereas it's the dowdy, less-pigmented flowers that are the most fragrant. This reflects a biochemical trade-off between the different terpenoid pathways, those leading towards colour and those towards odour[129]. Though, of course, on the 'have your cake and eat it' principle, centuries of human intervention in flower genetics and physiology have created a multitude of ornamentals that combine both traits.

Pigment	Leaf	Mature green fruit	Ripe fruit
Chlorophyll a+b	2200	106	0
β-carotene	625	10	123
Lutein	357	3	26
Neoxanthin	131	14	0
Violaxanthin	158	2	0
Lycopene	0	0	816
Phytoenes	0	0	144
Phytofluenes	0	0	224
Total carotenoids	1300	30	1300

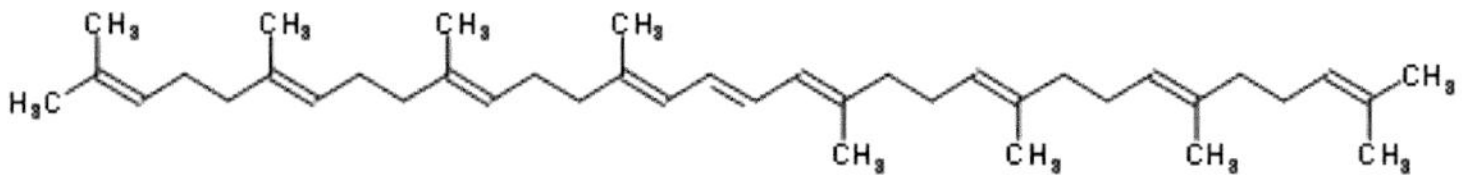

Phytoene

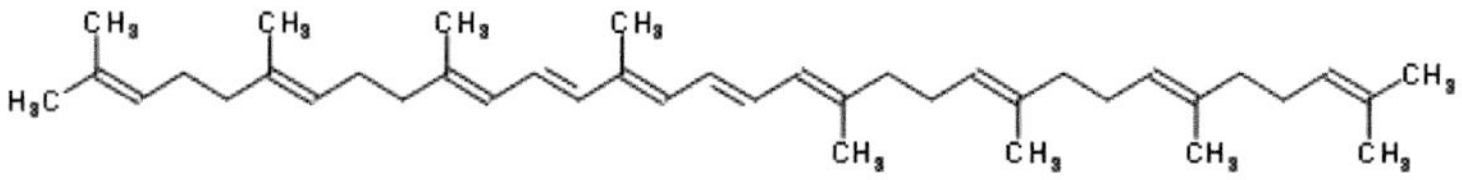

All-trans phytofluene

Amounts of chlorophylls and carotenoids (μg mg^{-1} dry weight) in tomato leaf, green fruit and red fruit[130]

The vivid colours and exotic perfumes of flowers are more than matched by the arresting visual, olfactory and dietary appeal of the fruits they develop into. We judge the ripeness of the fruits we're familiar with primarily by their colour (supermarkets know we buy with our eyes[131]) and secondarily by flavour, fragrance and texture. As with the evolution of the angiosperm flower, fruit attributes were subject to coevolutionary selection by the animals they attracted[132]. In particular, the biochemistry and regulation of ripening are distinguished by novel mechanisms for signalling to animal consumer/dispersers (including humans), and terpenoids are prominent factors in this relationship. We see this if we compare the pigment compositions of the leaf, fully-grown green fruit and mature red fruit of tomato[133]. The pigment profile of a green tomato resembles that of a leaf, but the amounts of chlorophylls and carotenoids per unit dry weight are orders of magnitude less - which tells us that a green tomato is mostly water. The total carotenoid content of the fruit at maturity is comparable with that of the leaf, but includes a range of novel pigments, chief among which are the red carotene lycopene and the deep orange β-carotene. A profile of xanthophylls, phytofluene and phytoene broadly similar to that of tomato accounts for most of the colours of citrus fruits. Varieties with red pulp, such as Cara-cara orange and Star Ruby grapefruit[134], also accumulate large amounts of lycopene[135]. For a fruit, the key to attracting dispersers is to be conspicuous, and this is achieved through not only colour but also morphology (water-powered hypertrophy in the case of tomato), texture and aroma[136].

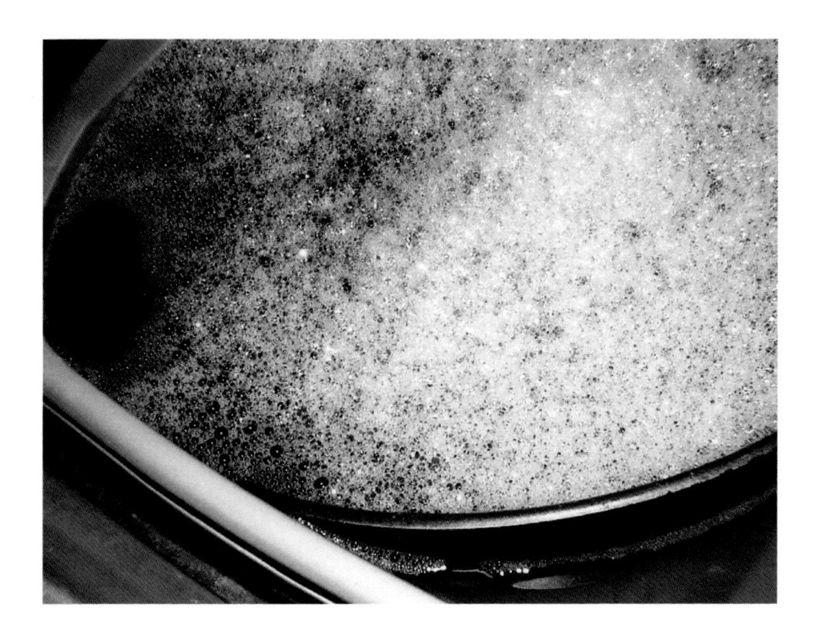

Tomato carotenoids are hydrophobic, preferring detergent to water[137]

Just as fruits lose their chlorophyll and become more colourful and conspicuous as they ripen, so too does the chlorophyll of foliage disappear during senescence. This unmasks carotenoids, whose colours become intensified by aggregation, further accumulation and chemical modification, giving the leaves of many trees in autumn their characteristic golden colour. The long hydrocarbon backbones of carotenoid molecules render them hydrophobic, strongly inclined to shun aqueous environments within cells and to bury themselves in membranes or lipid droplets. It's not surprising, therefore, that they are located more or less exclusively in the lipid-rich regions of plastids of senescing leaves or ripening fruits. It explains why the lycopene and carotene of tomato puree preferentially associate with oil globules, forming orange-red droplets on the surface of your ragu sauce when it cools – and why they concentrate in the detergent foam when the pan is washed. The carotenoids of ripening fruits, maturing petals and senescing leaves accumulate in globules, fibrils or crystal-like structures within the plastid as it becomes restructured from a chloroplast into a chromoplast[138]. Plastoglobules used to be thought of as middens, mere dumps of unwanted lipids and pigments, but now we know they are complex dynamic subcellular structures in their own right with important metabolic, developmental and adaptive functions and a long conserved evolutionary history[139]. The presence of carotenoid-rich globules helps minimise oxidation damage, thereby protecting the essential biochemical processes of ripening, maturation and senescence.

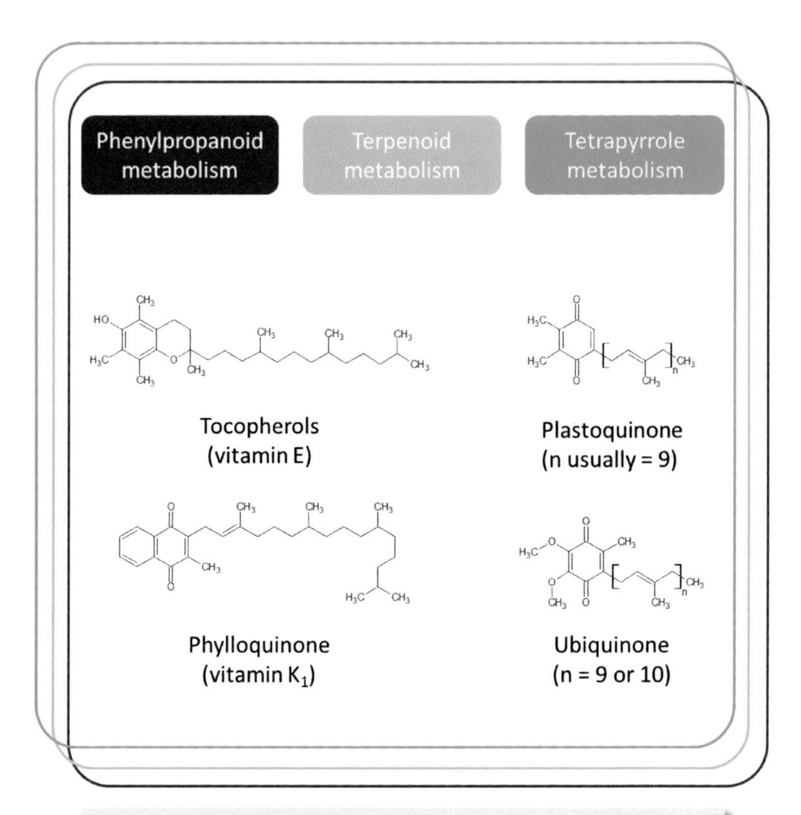

Prenylquinones are derivatives of all three little pigments

Now we come to the story of a special group of biomolecules that combines the chemistry of all three little pigments. Vitamin E was first described as a fat-soluble fertility factor in experiments on rats[140]. Subsequently it was identified as α-tocopherol, a component originally isolated from wheat germ oil, and was chemically synthesised in 1938 in the laboratory of Nobel prizewinner Paul Karrer[141]. Vitamin K_1 (phylloquinone) is a blood clotting factor and was characterised by Danish nutritionist Henrick Dam and American chemist Edward Doisy. Dam and Doisy won the Nobel Prize for medicine in 1943. Vitamins E and K_1 are essential for human health and, since only photosynthetic organisms can synthesise them, they must be obtained directly, or through the food chain, from plants. The common terpenoid precursor from which phylloquinone and tocopherols are synthesised is phytol. There are two possible sources of phytol: either de novo synthesis via the diterpenoid pathway, or as a salvage product of chlorophyll degradation[142]. In each case, the two oxygenated six-carbon rings to which the polyprenyl-derived side chain is attached are ultimately derived from the phenylpropanoid pathway (to be discussed further in the next section). A pair of phylloquinone molecules constitute part of the reaction centre of PSI, where they act as electron carriers. The family of prenylquinone compounds to which tocopherols and phylloquinone belong also includes plastoquinone, the mobile electron carrier in photosynthesis that links PSII with the cytochrome bf complex. The lipid globules of chloroplasts, which are in intimate contact with plastid membranes, are rich in prenylquinones and are the primary sites of biosynthesis and interconversion of these fat-soluble antioxidants[143]. Another family member is ubiquinone, structurally and functionally similar to plastoquinone. It's a respiratory electron carrier located in mitochondria, the ATP-generating organelles of animal and plant cells[144].

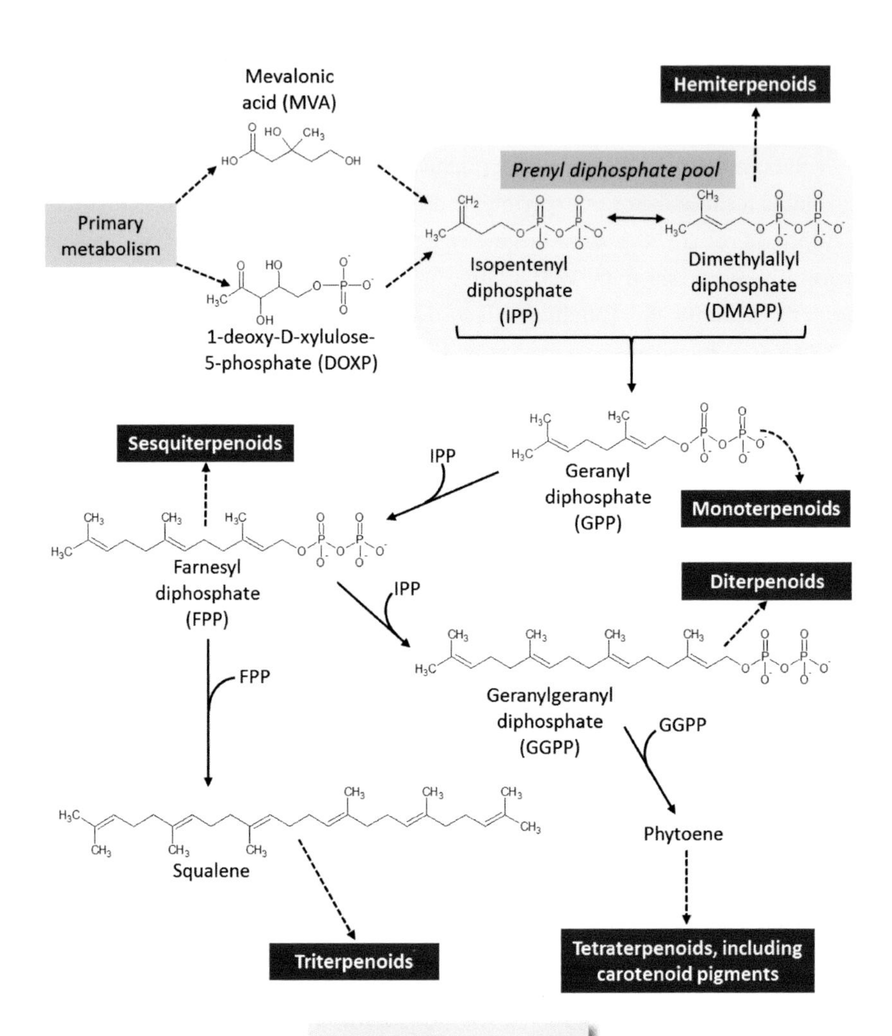

Terpenoid biosynthesis

Seen as a whole, the biochemical pathway leading to the synthesis of carotenoid pigments looks formidably complex, but is actually quite accessible once it's broken down into its major component processes[145]. The carbon skeletons of terpenoids originate in primary metabolism, specifically intermediates in the universal respiratory reactions that generate ATP from sugars and fats. Terpenoids are built from 5-carbon prenyl units, each supercharged with not one but two high-energy phosphate groups. One of the principal C5 precursors is IPP. IPP is made by two separate pathways, one derived from MVA and located in the cytosol, the other from DOXP in plastids[146]. The MVA route leads ultimately to phytosterols, sesquiterpenes and triterpenoids, whereas DOXP is the source of plastid terpenoids, including carotenoid pigments. The atoms of IPP are rearranged to make its isomer, DMAPP, and the equilibrium pool of IPP and DMAPP is the source of prenyl units for all terpenoids, from hemi- (C5, such as isoprene) to tetra- (C40 carotenoid pigments for example). The enzymic isomerisation of IPP, and the proportion of IPP to DMAPP in the pool, are critical points at which the amounts and types of terpenoid are regulated. Prenyl transferase enzymes sequentially add IPP molecules to a DMAPP primer, generating GPP (geranyl diphosphate, C10 leading to monoterpenoids), FPP (farnesyl diphosphate, C15 source of sesquiterpenoids) and GGPP (geranylgeranyl diphosphate, precursor of C20 diterpenoids including phytol). The activities of prenyl transferases lead as far as C20 terpenoids, most of which have conjugated bonding systems in the trans configuration. Larger molecules, including carotenoid pigments, are produced by joining mono-, sesqui- and di-terpenoids head-to-head, catalysed by synthase enzymes[147]. The triterpenoid squalene, the precursor of a range of steroids, is made from two FPP molecules. Phytoene, the tetraterpenoid precursor of the carotenoids, is the product of two GGPPs joined together. The enzymes responsible, squalene synthase and phytoene synthase respectively, catalyse a complex series of rearrangements necessary to align the head-end (C-1) carbon atoms of the two prenyl diphosphate precursors together.

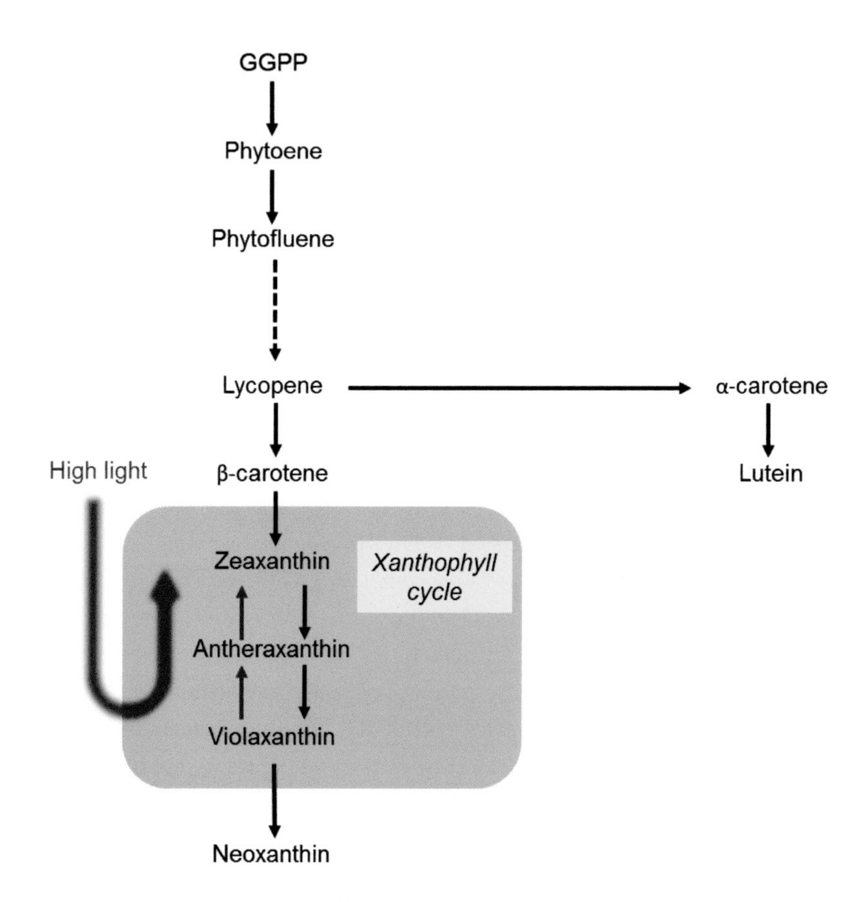

Biosynthesis of carotenoid pigments, and the photoregulatory xanthophyll cycle

Carotenoids are derived from phytoene by a series of desaturation reactions (which introduce additional double bonds to the molecule) and cyclisations, which form terminal rings. The pathway branches at lycopene. One way leads to lutein via α-carotene. The other is the route through β-carotene to most of the xanthophylls of green cells, culminating in neoxanthin. We've seen that light can be harmful to plants if the photon energy absorbed by chlorophyll is greater than the capacity of the plant to use it. Over-excited chlorophyll is a potent photosensitiser, which can cause cell damage through ROS and free radical cascades. Plants are equipped with several mechanisms to dissipate (quench) excessive energy from chlorophyll, among the most important of which are the biosynthesis and antioxidant reactions of carotenoids. Through the reactivity of their extended conjugated bonding systems, carotenoids are able to act as quenchers by accepting excitation energy from excited-state chlorophyll, thereby blocking the generation of ROS and free radicals. But the pathway of xanthophyll biosynthesis is more than simply a source of essential antioxidant pigments and derivatives; the enzymic reactions that interconvert zeaxanthin, antheraxanthin and violaxanthin are themselves quenching regulators. When light levels are low, violaxanthin functions as a photosynthetic antenna pigment by transferring energy to chlorophyll a. But under high light, water-splitting associated with PSII is intensified, leading to acidification within the thylakoid membrane and direct activation of the de-epoxidase enzyme which rapidly converts violaxanthin to antheraxanthin and thence to zeaxanthin, a quencher of excited-state chlorophyll. Such a so-called non-photochemical quenching mechanism directly prevents the buildup of excess excitation energy within light-harvesting antennae by harmlessly dissipating it in the form of heat. Because carotenoids carry out such essential protective functions in photosynthetic tissues, mutations that block terpenoid biosynthesis usually result in the formation of lethal concentrations of ROS under high light intensity. And chemicals that interfere with carotenoid metabolism have potent herbicidal properties.

Baltic amber[148]

Kauri gum[149]

The main biosynthetic routes from the prenyl diphosphate pool to carotenes and xanthophylls have many branch lines, leading to a much greater variety of terpenoids in plants than is produced by either animals or microbes. An extraordinary example of the antiquity and pervasiveness of terpenoids is the group of C30 derivatives called hopanoids, which are said to be the most ancient[150] and abundant[151] natural products on earth. In plants, the large-scale accumulation, emission or secretion of terpenoids is almost always associated with anatomical specialisation. Structures such as resin ducts, secretory cavities, epidermal hairs and blisters are sources of defensive products, essential oils, triterpenoid surface waxes, latex, rubber, gums, saps and mucilages. In the context of pigments, amber is a special case, being a complex terpenoid derivative that gives its name to a familiar, and much sought-after coloured material of plant origin. Amber is the fossilised member of a family of products derived from the resinous secretions of (mostly) coniferous trees, a family that includes resin, rosin, copal, Mayan pom and New Zealand kauri gum. Amber comes in a variety of colours in addition to the golden hue seen in traffic lights – everything from red through blue to black. The largest known single deposit of amber is in the Baltic region, where it is estimated that 100,000 tons of conifer resin was fossilised in the Eocene epoch, more than 40 million years ago[152]. Together with its subfossilised relative copal[153], amber is famous for high-fidelity preservation of trapped small, soft bodied terrestrial organisms, such as arthropods, and delicate plant parts, notably flowers and leaves, giving unique insights into the evolution of morphology, anatomy and behaviour, including herbivory, parasitism, pollination, mimicry and mating[154]. Solvent extraction of amber yields a range of monoterpenoids (C10) such as bornene and camphor, and diterpenoids (C20) such as abietic acid. The insoluble residue consists of large cross-linked molecules formed by polymerisation, cyclisation and photo-oxidation of resin terpenoids during fossilisation. For more detail about amber chemistry than you could possibly wish for, the review by Vávra[155] is recommended.

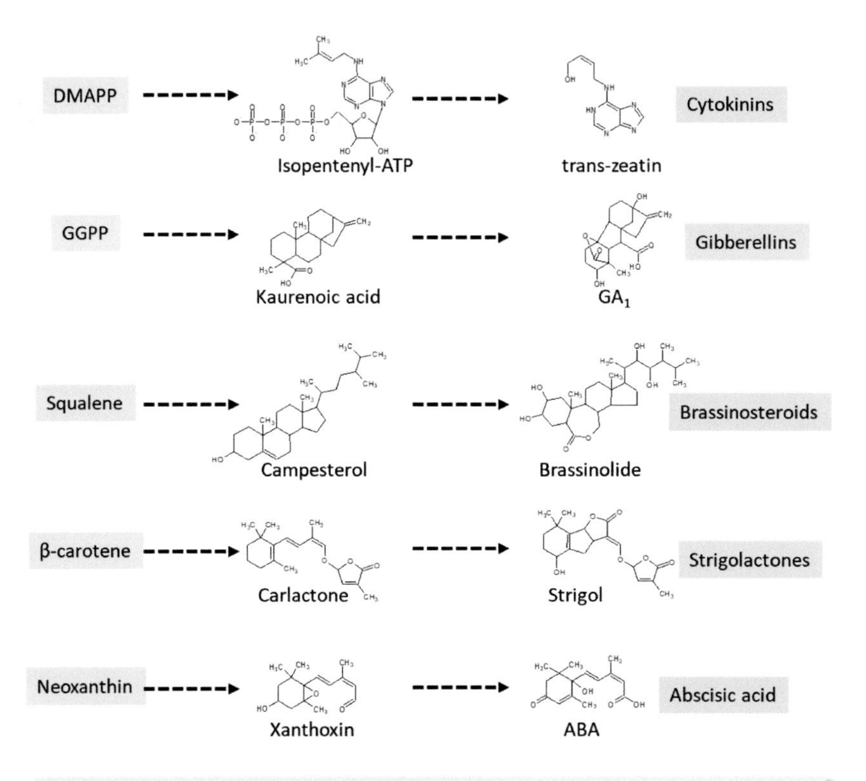

Terpenoid biosynthesis is the source of most of the main classes of plant hormone

An essential function of the pathway of terpenoid biosynthesis is to provide precursors for the synthesis of plant hormones[156], which in turn are the controllers of growth, form and function (including pigmentation). Cytokinins are prenylated derivatives of ATP, the cell's universal energy currency for biochemical transactions. The addition of the C5 group to ATP is carried out by the enzyme isopentenyl transferase (IPT). Chlorophyll breakdown during leaf senescence is much delayed in plants overexpressing IPT[157]. As well as their role in regulating de-greening and senescence, cytokinins are active in branching and cell division. Gibberellins (GAs) are diterpenoid derivatives, synthesised from kaurenoic acid, a product of GGPP cyclisation[158]. GAs have a wide range of hormonal effects, regulating seed germination, bud and seed dormancy and flowering. A function of particular significance for Green Politics, to be described in more detail subsequently, is the control of plant height. Brassinosteroids and strigolactones are recently described cyclic terpenoids derived from squalene and β-carotene respectively, which are active in various aspects of plant development[159]. Campesterol, the intermediate in brassinosteroid synthesis, is representative of a whole family of membrane-associated sterols (with a characteristic three C6 plus one C5 ring structure) that includes cholesterol and the insect moulting hormone ecdysone. Hobson[160] was the first to show that morphogenesis of insect larvae is dependent on plant sterols in the diet. Some plants accumulate ecdysone-like sterols that disrupt the development of infesting insects[161]. Abscisic acid (ABA) is a hormone that in many ways acts as a GA antagonist; for example, it promotes dormancy while GA inhibits it[162]. ABA also has a central role in plant responses to water availability[163]. Under drought conditions, ABA biosynthesis is stimulated by a long-distance hydraulic signal, and rehydration rapidly leads to ABA breakdown. As well as enhancing the intrinsic tolerance of tissues to water limitation, ABA impedes dehydration by regulating the opening and closing of stomata. ABA is an apocarotenoid, formed by asymmetrical oxidative cleavage of neoxanthin. Apocarotenoids, the products of an extensive family of carotenoid cleavage dioxygenase enzymes, are widely distributed in living organisms, frequently performing important regulatory, receptor and signalling functions[164].

'Miracle rice': Dee-geo-woo-gen (DGWG), source of dwarfing genes crossed into tall variety Peta to produce the Green Revolution rice variety IR8[165]

Gregor Mendel, the Father of Genetics, studied the inheritance of seven different traits in peas, one of which was plant stature. He found that height was inherited as what today we would call a single-gene character. Genetic variation for stature is widespread in crop species and it was exploitation of genetic diversity for height that became the basis of the Green Revolution[166]. By introducing semi-dwarf varieties of rice and wheat, Peter Jennings, Henry Beachell, Gurdev Singh Khush, Orville Vogel and other plant breeders (heroic but largely unknown benefactors of humanity) were able greatly to improve grain yields. Plants selected for reduced stature are less likely to lodge (that is, to fall over in the field), they have increased responsiveness to fertiliser, and they allocate more photosynthate to grain and less to stem[167]. The most important genes exploited in breeding modern short-stemmed, high-yielding varieties are those concerned with synthesis of or sensitivity to GAs. Many semi-dwarf rice varieties are deficient in GA because a mutation in the gene *sd1* results in a defective oxidase enzyme, thereby causing a blockage in the terpenoid pathway leading from kaurenoic acid[168]. Short-straw wheat is different. In this species, it's variants of the *Rht* gene that are responsible for the dwarf trait[169]. Mutations in *Rht* reduce the growth response to GA by interfering with the hormone perception mechanism. To be able to trace the chain of cause and effect from the molecular detail of differences in DNA sequence through the biochemistry of hormone synthesis and function to the growth and yield of a food plant and out into the real world of agriculture and famine relief is what thrills and inspires a crop scientist (well, like me, actually). But the Green Revolution remains contentious and its achievements, motivations and costs (sociopolitical, economic and ecological), are open to criticism[170]. It's a recurring question: given we can't just do nothing when faced with the Malthusian predicament, what should we do for the best?

The gold vexation

Too much of a good thing: among the fossil hominids described by Richard Leakey and colleagues is a partial skeleton of *Homo erectus* from Kenya showing pathological changes consistent with chronic vitamin A poisoning. It was suggested that hypervitaminosis symptoms were the result of a change in the dietary habits of *H. erectus* to include a high intake of animal liver[171]. There is a long history of vitamin A poisoning among arctic travellers and fishermen who consumed the livers of polar bears, seals, husky dogs or flatfish[172].

Generally speaking, however, hypervitaminosis A is a rare condition. On the other hand, UNICEF estimates that more than 140 million pre-school children and more than seven million pregnant women, are seriously vitamin A *deficient,* precipitating over a million child deaths each year. Of the 118 countries affected, most are in Asia, sub-Saharan Africa and South and Central America. Vitamin A deficiency is a leading cause of child blindness in developing countries. It also limits growth, weakens the body's immune system and increases mortality. The condition and its treatment have been understood at least as far back as the time of the classical Greeks, and possibly even the ancient Egyptians[173]. Consume liver (or fish liver oil) is the answer. But what if, as throughout the developing world, these remedies are of limited availability, too expensive, or precluded culturally by adherence to a vegetarian diet?

The Golden Rice project[174] was born as a response to this predicament. Rice grains, the main food source in many of the regions with endemic vitamin D deficiency, are low in β-carotene, a precursor of vitamin A synthesis. But rice leaves, like all green plant tissues, make plenty of β-carotene. There's evidently a gap somewhere in the terpenoid synthesis pathway of rice grains, the consequence of genes being disabled during seed development. It turned out this blockage could be overcome by introducing genes encoding two enzymes of carotenoid biosynthesis: the synthetase that converts endogenous GGPP into phytoene; and a bacterial enzyme that introduces four double bonds into phytoene, completing the metabolic sequence leading to carotenes[175]. It took years to get the system working with high efficiency but now there are varieties of rice, biofortified with golden carotenoids, to levels in excess of 30 $\mu g\ g^{-1}$ fresh weight, that amply meet the recommended dietary daily allowance for adequate conversion to vitamin A. By 'introducing genes' I mean, of course, using transgenics, which pitches Golden Rice straight into the ongoing stand-off between biotechnology and the Green Movement[176]. This issue is complex, fast-moving and won't be analysed here. Suffice to say that the scientists who conceived and realised the Golden Rice project really should be household names, respected for their brilliant work: Ingo Potrykus, Peter Beyer, Adrian Dubock, Peter Bramley. They were motivated to apply their deep knowledge and expertise to solve a humanitarian problem. What have their extremist opponents contributed? Tearing down fences and tearing up plants. And the politics of this is getting uglier. It's now being suggested that online information warfare, conducted by state-sponsored hackers, is driving even further the wedge between the scientists and environmentalists who ought to be on the same side[177].

Meanwhile, the vitamin deficient wait, suffer and die.

NOTES AND SOURCES

[102] Public domain image. Carotenoids account for the golden colour of persimmon (Ebert G, Gross J. 1985. Carotenoid changes in the peel of ripening persimmon (*Diospyros kaki*) cv Triumph. Phytochemistry 24:29-32).
[103] Image by http://freebigpictures.com/mountain-pictures/misty-hills/ [accessed 25 May 2018].
[104] Galston AW, Sharkey TD. 1998. Frits Warmolt Went. May 18, 1903 - May 1, 1990. Biographical Memoirs, National Academy of Sciences of the United States of America 74: 348-363
[105] Went FW. 1960. Blue hazes in the atmosphere. Nature 187: 641-643.
[106] Harrison SP, Morfopoulos C, Dani KGS, Prentice IC, Arneth A, Atwell BJ, Barkley MP, Leishman MR, Loreto F, Medlyn BE, Niinemets Ü, Possell M, Peñuelas J, Wright IJ. 2013. Volatile isoprenoid emissions from plastid to planet. New Phytologist 197: 49–57.
[107] For information on all things carotenoid, there's the Carotenoid Society: www.carotenoidsociety.org/about-carotenoids [accessed 14 May 2018].
[108] Sourkes TL. 2009. The discovery and early history of carotene. Bulletin for the History of Chemistry 34: 32-38.
[109] Ohmiya A. 2011. Diversity of carotenoid composition in flower petals. Japan Agricultural Research Quarterly JARQ 45: 163-171.
[110] Harman D. 1956. Aging: a theory based on free radical and radiation chemistry. Journal of Gerontology 11: 298–300; Halliwell B, Gutteridge JMC. 2015. *Free Radicals in Biology and Medicine* (5th edition). Oxford: University Press.
[111] McGuire M, Beerman KA. 2012. *Nutritional Sciences: From Fundamentals to Food* (3rd edition). Belmont: Wadsworth Cengage Learning.
[112] Olson VA, Owens IPF. 1998. Costly sexual signals: are carotenoids rare, risky or required? Trends in Ecology and Evolution 13: 510–515.
[113] Jones RL, Ougham H, Thomas H, Waaland S. 2012. *Molecular Life of Plants*. Chichester: Wiley-ASPB.
[114] There are many examples of pigment separation by chromatography out there. I can't remember, and haven't been able to find, where I got this particularly nice one. If it's yours, I'll be happy to acknowledge in a future edition.
[115] Ohmiya (2011).
[116] Takaichi S. 2011. Carotenoids in algae: distributions, biosyntheses and functions. Marine Drugs 9: 1101–1118.
[117] Xiong J, Bauer CE. 2002. Complex evolution of photosynthesis. Annual Review of Plant Biology 53: 503-521.
[118] Xiong J. 2007. Photosynthesis: what color was its origin? Genome Biology 7: 245.
[119] Ohmiya (2011).
[120] Telfer A. 2002. What is beta-carotene doing in the photosystem II reaction centre? Philosophical Transactions of the Royal Society B: Biological Sciences 357: 1431–1470.
[121] Karrer P, Helfenstein A, Wehrli H, Wettstein A. 1930. Pflanzenfarbstoffe XXV. Über die Konstitution des Lycopins und Carotins. Helvetica Chimica Acta 13: 1084–1099.

[122] Pogson B, McDonald KA, Truong M, Britton G, DellaPenna D. 1996. Arabidopsis carotenoid mutants demonstrate that lutein is not essential for photosynthesis in higher plants. Plant Cell 8: 1627-1639.
[123] Takaichi S, Mirauro M. 1998. Distribution and geometric isomerism of neoxanthin in oxygenic phototrophs: 9-cis, a sole molecular form. Plant and Cell Physiology 39: 968-977.
[124] Li X, Lu M, Tang D, Shi Y. 2015. Composition of carotenoids and flavonoids in Narcissus cultivars and their relationship with flower color. PLoS ONE 10(11): e0142074.
[125] Valadon LRG, Mummery RS. 1968. Carotenoids in floral parts of a narcissus, a daffodil and a tulip. Biochemical Journal 106: 479–484.; Li et al (2015).
[126] Ohmiya (2011).
[127] Plackett AR, Di Stilio VS, Langdale JA. 2015. Ferns: the missing link in shoot evolution and development. Frontiers in Plant Science 6: 972.
[128] Thomas H, Sadras VO. 2000. The capture and gratuitous disposal of resources by plants. Functional Ecology 15: 3-12.
[129] Borghi M, Fernie AR, Schiestl FP, Bouwmeester HJ. 2017. The sexual advantage of looking, smelling, and tasting good: the metabolic network that produces signals for pollinators. Trends in Plant Science 22: 338-350.
[130] Enfissi EMA, Nogueira M, Bramley PM, Fraser PD. 2017. The regulation of carotenoid formation in tomato fruit. Plant Journal 89: 774–788.
[131] A tomato is a marketing manager's dream come true: a simple sphere, a bright primary colour, an acknowledged functional food reputation.
[132] Schaefer HM, Schaefer V, Vorobyev M. 2007. Are fruit colors adapted to consumer vision and birds equally efficient in detecting colorful signals? American Naturalist 169: S159-169.
[133] Enfissi et al. (2017).
[134] Cara-cara is a navel and not to be confused with blood orange, where the red pigment is anthocyanin.
[135] Lado J, Zacarías L, Gurrea A, Page A, Stead A, Rodrigo MJ. 2015. Exploring the diversity in *Citrus* fruit colouration to decipher the relationship between plastid ultrastructure and carotenoid composition. Planta 242: 645-661.
[136] Valido A, Schaefer HM, Jordano P. 2011. Colour, design and reward: phenotypic integration of fleshy fruit displays. Journal of Evolutionary Biology 24: 751–760.
[137] Pan hand-washed by the author.
[138] Tevini, M, Steinmüller D. 1985. Composition and function of plastoglobuli. Planta 163: 91-96; Kilcrease J, Collins AM, Richins RD, O'Connell MA. 2013. Multiple microscopic approaches demonstrate linkage between chromoplast architecture and carotenoid composition in diverse *Capsicum annuum* fruit. Plant Journal 76: 1074-1083.
[139] van Wijk KJ, Kessler F. 2017. Plastoglobuli: plastid microcompartments with integrated functions in metabolism, plastid developmental transitions, and environmental adaptation. Annual Review of Plant Biology 68: 253-289; Lohscheider JN, Bártulos CR. 2016. Plastoglobules in algae: a comprehensive comparative study of the presence of major structural and functional components in complex plastids. Marine Genomics 28: 127-136.
[140] Evans HM, Bishop KS. 1922. On the existence of a hitherto unrecognized dietary factor essential for reproduction. Science 56: 650–651.
[141] Karrer P, Fritzsche H, Ringier BH, Salomon H. 1938. Synthesis of α-tocopherol (vitamin E). Nature. 141: 1057.

[142] Mach J. 2015. Phytol from degradation of chlorophyll feeds biosynthesis of tocopherols. Plant Cell 27: 2676.
[143] Spicher L, Kessler F. 2015. Unexpected roles of plastoglobules (plastid lipid droplets) in vitamin K_1 and E metabolism. Current Opinion in Plant Biology 25: 123-129.
[144] Liu M, Lu S. 2016. Plastoquinone and ubiquinone in plants: biosynthesis, physiological function and metabolic engineering. Frontiers in Plant Science 7: 1898.
[145] Jones et al. (2012).
[146] Lange BM, Rujan T, Martin W, Croteau R. 2000. Isoprenoid biosynthesis: the evolution of two ancient and distinct pathways across genomes. Proceedings of the National Academy of Sciences 97: 13172-13177.
[147] Boutanaev AM, Moses T, Zi J, Nelson DR, Mugford ST, Peters RJ, Osbourn A. 2015. Investigation of terpene diversification across multiple sequenced plant genomes. Proceedings of the National Academy of Sciences 112: E81-98; Llorente B, Martinez-Garcia JF, Stange C, Rodriguez-Concepcion M. 2017. Illuminating colors: regulation of carotenoid biosynthesis and accumulation by light. Current Opinion in Plant Biology 37: 49-55.
[148] CC0 image by Michal Kosior, from https://commons.wikimedia.org/wiki/File:Colours_of_Baltic_Amber.jpg [accessed 26 May 2018].
[149] With thanks to Nancy Devlin.
[150] Brocks J, Logan G, Buick R, Summons R. 1999. Archean molecular fossils and the early rise of eukaryotes. Science 285: 1033–1036.
[151] Ourisson G, Albrecht P. 1992. Hopanoids. 1. Geohopanoids: the most abundant natural products on earth? Accounts of Chemical Research 25: 398–402.

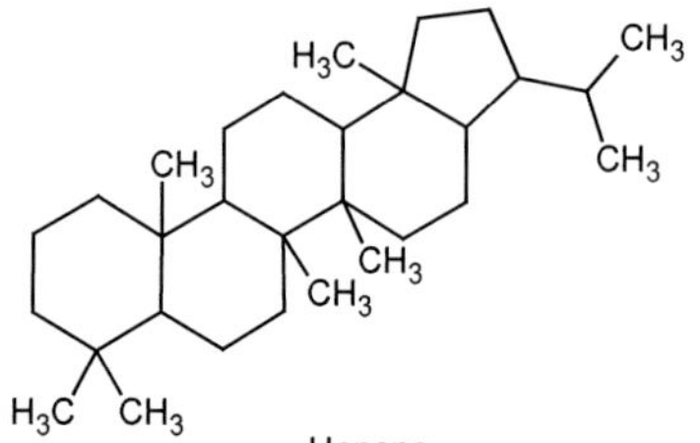

Hopane

[152] Wolfe AP, Tappert R, Muehlenbachs K, Boudreau M, McKellar RC, Basinger JF, Garrett A. 2009. A new proposal concerning the botanical origin of Baltic amber. Proceedings of the Royal Society B: Biological Sciences. 276: 3403–3412.
[153] Case RJ, Tucker AO, Maciarello MJ, Wheeler KA. 2003. Chemistry and ethnobotany of commercial incense copals: copal blanco, copal oro, and copal negro, of North America. Economic Botany 57: 189–202.
[154] McCoy VE, Soriano C, Gabbott SE. 2018. A review of preservational variation of fossil inclusions in amber of different chemical groups. Earth and Environmental Science Transactions of The Royal Society of Edinburgh 107: 203-211.
[155] Vávra N. 2009. The chemistry of amber - facts, findings and opinions. Annalen des Naturhistorischen Museums in Wien. Serie A für Mineralogie und Petrographie, Geologie und Paläontologie, Anthropologie und Prähistorie. January 1: 445-473.
[156] Jones et al. (2012).

[157] Gan S, Amasino RM. 1995. Inhibition of leaf senescence by autoregulated production of cytokinin. Science 270: 1986-1988.
[158] Zi J, Mafu S, Peters RJ. 2014. To gibberellins and beyond! Surveying the evolution of (di) terpenoid metabolism. Annual Review of Plant Biology. 65: 259-286.
[159] Fridman Y, Savaldi-Goldstein S. 2013. Brassinosteroids in growth control: how, when and where. Plant Science 209: 24-31; Marzec M. 2016. Perception and signaling of strigolactones. Frontiers in Plant Science 7:1260.
[160] Hobson RP. 1935. On a fat-soluble growth factor required by blowfly larvae. II. Identity of the growth factor with cholesterol. Biochemistry Journal 29: 2023-2026.
[161] Adler JH, Grebenok RJ. 1995. Biosynthesis and distribution of insect-molting hormones in plants—a review. Lipids 30: 257–262.

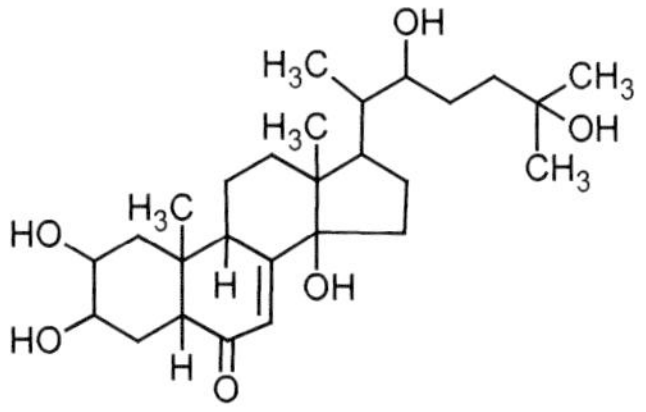

Ecdysone

[162] Cooke JE, Eriksson ME, Junttila O. 2012. The dynamic nature of bud dormancy in trees: environmental control and molecular mechanisms. Plant Cell and Environment 35: 1707–1728.
[163] Seki M, Umezawa T, Urano K, Shinozaki K. 2007. Regulatory metabolic networks in drought stress responses. Current Opinion in Plant Biology 10: 296-302.
[164] Walter MH, Strack D. 2011. Carotenoids and their cleavage products: biosynthesis and functions. Natural Product Reports 28: 663–692.
[165] Nagai K, Hirano K, Angeles-Shim RB, Ashikari M. 2018. Breeding applications and molecular basis of semi-dwarfism in rice. In: Sasaki T, Ashikari M. (eds) *Rice Genomics, Genetics and Breeding*. Singapore: Springer, pp 155-176. Image under CC BY-NC-SA 2.0 from Baroña-Edra L. 2013. Upon the 100,000th cross. Rice Today 12: 17-18.
[166] Hedden P. 2003. The genes of the Green Revolution. Trends in Genetics. 19: 5-9.
[167] https://livinghistoryfarm.org/farminginthe50s/crops_17.html [accessed 25 June 2018].
[168] Sasaki A, Ashikari M, Ueguchi-Tanaka M, Itoh H, Nishimura A, Swapan D, Ishiyama K, Saito T, Kobayashi M, Khush GS, Kitano H. 2002. Green revolution: a mutant gibberellin-synthesis gene in rice. Nature 416: 701.
[169] Würschum T, Liu G, Boeven PH, Longin CF, Mirdita V, Kazman E, Zhao Y, Reif JC. 2018. Exploiting the Rht portfolio for hybrid wheat breeding. Theoretical and Applied Genetics 131: 1433–1442.
[170] For example, Gan E. 2017. An unintended race: miracle rice and the Green Revolution. Environmental Philosophy 14: 61-81.
[171] Walker A, Zimmerman MR, Leakey RE. 1982. A possible case of hypervitaminosis A in *Homo erectus*. Nature 296:248.
[172] Rodahl K, Moore T. 1943. The vitamin A content and toxicity of bear and seal liver. Biochemical Journal 37: 166-168.
[173] Wolf G. 1996. A history of vitamin A and retinoids. FASEB Journal 10: 1102-1107.

[174] Some useful sources: www.goldenrice.org/index.php [accessed 23 May 2018]; https://geneticliteracyproject.org/2018/02/13/golden-rice-gmo-crop-greenpeace-hates-and-humanitarians-love/ [accessed 23 May 2018]; Eisenstein M. 2014. Biotechnology: against the grain. Nature 514: S55-57.
[175] Schaub P, Al-Babili S, Drake R, Beyer P. 2005. Why is Golden Rice golden (yellow) instead of red? Plant Physiology 138: 441-450.
[176] Dubock A. 2017. An overview of agriculture, nutrition and fortification, supplementation and biofortification: Golden Rice as an example for enhancing micronutrient intake. Agriculture and Food Security 6: 59.
[177] Dorius SF, Lawrence-Dill CJ. 2018. Sowing the seeds of skepticism: Russian State News and anti-GMO sentiment. Open Science Framework. March 6. osf.io/v27yh.

4. PURPLE

Concerning phenylpropanoids, a chemical family that includes a diversity of pigments

If Bacchus ever had a colour he could claim for his own, it should surely be the shade of tannin on drunken lips, of John Keat's 'purple-stained mouth', or perhaps even of Homer's dangerously wine-dark sea.

Victoria Finlay (*Colour: Travels Through the Paintbox*)

Rainbow corn: genetic variants in phenylpropanoid pigmentation[178]

Cytogeneticists are different[179]. I've worked (and published) with them for many years and know from close observation that it takes a distinctive kind of scientific imagination to be a cytogeneticist. You have to posses a particular creativity to be able to infer the nature and development of an entire organism from the strange hieroglyphic patterns formed by chromosomes during the cell cycle. The doyenne of cytogeneticists was Barbara McClintock (1902-92), one of the very few plant scientist Nobel laureates[180]. In just 43 publications she virtually wrote the book on chromosome genetics. Among her many great accomplishments was the discovery of 'controlling elements' - what we now call transposons - by analysing the pattern of pigmentation in maize kernels. The figure of the 'hermit scientist', a solitary genius dismissed by the scientific and social establishment as a crank, is pretty much a myth[181]. McClintock's style was certainly solitary and independent, and so deep and unprecedented were her accomplishments in cytogenetics that she almost conforms to the stereotype. She even ceased publishing her work for some years from 1953 because of the baffled, and even hostile, reaction from the community[182]. Now, however, she is recognised as a true giant of modern biology. It has become a recurring theme in this book that some of the highest achievements in science feature one aspect or another of plant pigmentation. In the case of McClintock's maize, the coloured spots, streaks and sectors in kernel lineages on the cob are products of the genetic control of phenylpropanoid metabolism. I'll have more to say on the subject of pigments in cereal grains subsequently. As for the detailed cytogenetics that led McClintock to the concept of transposition from observations on maize pigmentation variants, this is a challenging subject which isn't appropriate for discussion here. But fortunately my friend and colleague Neil Jones (another fine cytogeneticist) has published a clear summary of the events surrounding the 'genetic earthquake' of 1944 which resulted in the discovery of the *Ac-Ds* system of mobile genetic elements[183]: highly recommended if you want an insight into the unique mind of Barbara McClintock.

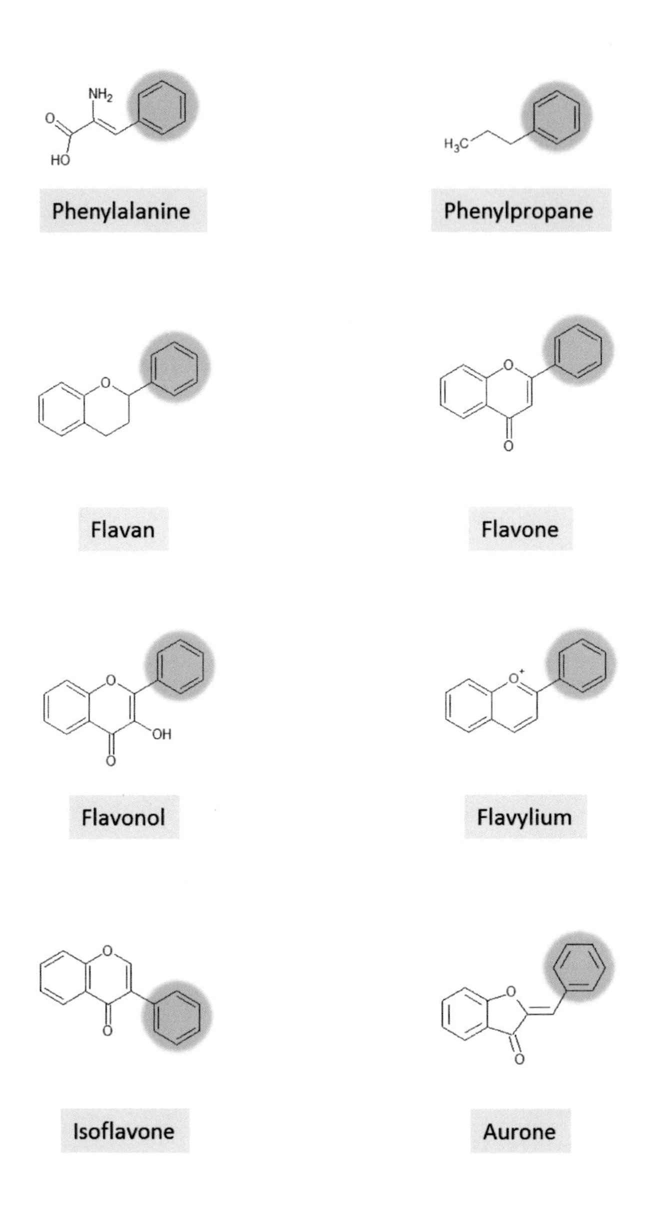

Phenylpropanoids, named for their structural relationship to phenylpropane, are biosynthesised from the amino acid phenylalanine

Phenylpropanoids get their not-very-attractive name from the relationship of their chemical structures to phenylpropane, which consists of a C_6 aromatic (phenyl) ring with a C_3 (propane) side-chain[184]. Phenylpropanoid pigments are members of a highly diverse class of non-structural compounds which also includes tannins, volatiles and defence compounds. Among the structural phenylpropanoids is lignin, the second most abundant natural polymer (after cellulose), responsible for the rigidity and resilience of wood and other fibrous tissues. Most phenylpropanoids are metabolically derived from the amino acid phenylalanine, with tyrosine and tryptophan also acting as precursors to certain related secondary compounds[185]. The bewildering variations on the phenylpropanoid theme show plants at their most chemically inventive, and have long subjected non-specialists in secondary metabolism to terminological vexation. Let's try to set out the essentials. The main pigment group we'll be discussing is the flavonoids, based on the flavan structural skeleton[186]. This consists of two C_6 phenyl rings (one of which relates to the parent phenylpropane) and an intermediate C_3 oxygen-containing heterocyclic ring[187]. The 'a' of flavan becomes an 'o' in flavone, to signify the presence of a keto group (oxygen double-bonded to the C_3 ring). Flavonoids, and their isoflavonoid relatives[188], are ketones. Aurones, responsible for the yellow-gold colours of cosmos and snapdragon flowers, have a 5-membered intermediate heterocyclic ring[189]. Anthocyanin pigments, with which we will be much concerned, are flavonoid derivatives based on the flavylium ion, a modified flavan skeleton that lacks the keto group and carries a positively charged oxygen. The illustrious Robert Boyle, of Royal Society fame, is credited with some of the earliest studies of flavonoid pigments (*Experiments and Considerations Touching Colours*, 1664[190]), describing the effects of acids and bases on the colour of extracts from flowers and other plant tissues. Once again, the Nobel prize enters the story of plant pigments: Albert Szent-Gyorgyi (discoverer of vitamin C, Nobel laureate 1936) observed the anti-scorbutic properties of flavonoids (they were originally called vitamin P) in preparations from paprika and citrus peel[191]. And here comes another Nobel prizewinner: the molecular basis of colour in anthocyanidins and their derivatives was first explained by Linus Pauling[192].

Because lignified tissue decays slowly, it sequesters carbon (or so it's hoped)

The capacity to make a huge range of phenylpropanoids - estimated to number at least 8,000 different compounds - with a variety of structural and physiological roles is one of the characteristics that facilitated the evolution of terrestrial plants from their aquatic ancestors[193]. Lignins stiffen cell walls, thereby holding land plants upright in the absence of the buoyant support provided by water[194]. A diversity of non-structural phenylpropanoids contribute to attributes essential for survival such as hardiness, colour, taste, odour and defence against biotic and abiotic stresses. It's estimated that about 20% of the carbon fixed by plants flows through the aromatic amino acid pathway and into phenylpropanoids, which in turn account for about 40% of organic carbon circulating in the biosphere. The biodegradation of relatively refractory lignins, tannins and suchlike macromolecules is a rate-limiting step in the recycling of organic carbon into CO_2, which makes sequestration in phenylpropanoid-rich biomass such as trees of central interest in carbon capture and greenhouse gas mitigation strategies. The flavonoids, comprising an estimated 4,500 different chemical structures, are the most diverse group of plant phenylpropanoids[195]. They include: anthocyanins (responsible for the pink/blue/purple/red pigmentation in many vegetative tissues, flowers and fruits); proanthocyanidins or condensed tannins (antifeedants and wood protectants); and isoflavonoids (defensive products and signalling molecules).

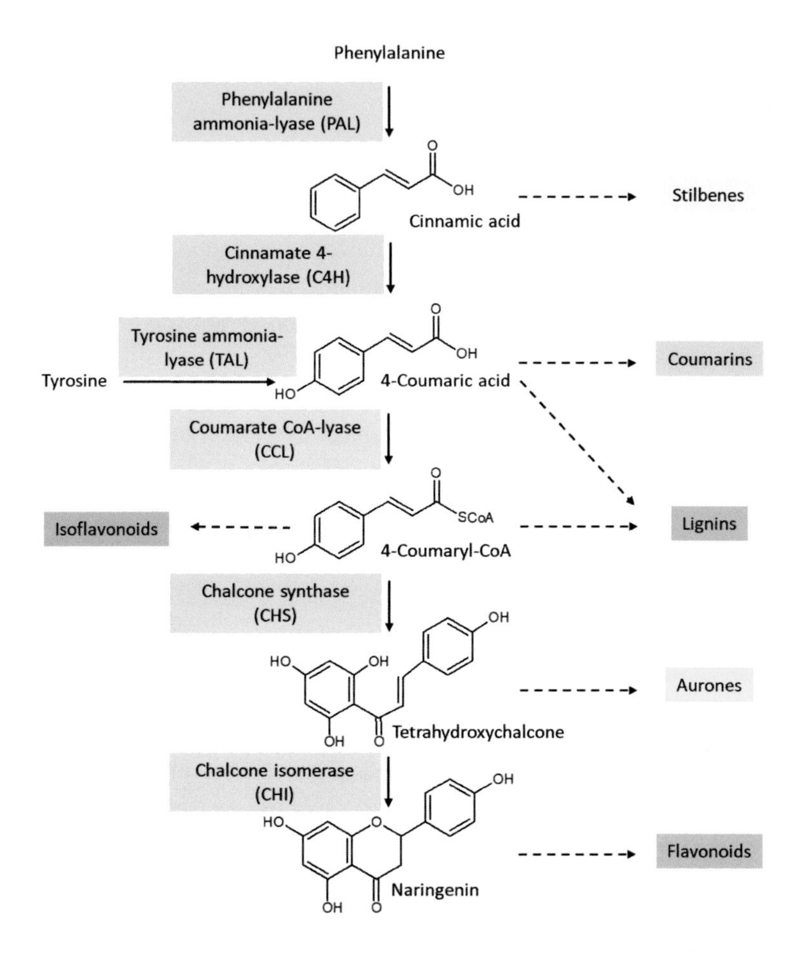

Enzymes and intermediates in the early stages of phenylpropanoid biosynthesis from phenylalanine

The detailed pathways of phenylpropanoid metabolism are fiendishly complicated[196], and so I will stick to the salient points. Phenylalanine is the link between, on the one hand, primary metabolism and protein synthesis and, on the other, the secondary pathways of phenylpropanoid biosynthesis. Phenylalanine, tyrosine and tryptophan[197] are also sources of other important secondary metabolites, including bioactive alkaloids and auxin growth hormones. PAL is the enzyme that stands at the entrance to phenylpropanoid metabolism[198]. It removes the nitrogenous amine group of phenylalanine as ammonia, yielding cinnamic acid, the parent compound from which most members of the phenylpropanoid family are derived[199]. The phenyl ring of cinnamate is hydroxylated by the enzyme C4H to make 4-coumarate. Cinnamate is the precursor of stilbenes, and 4-coumarate is the source of coumarins. In a complex series of biochemical reactions, three critical enzymes, CCL, CHS and CHI, convert 4-coumarate into naringenin via 4-coumaryl-CoA, the metabolic origin of isoflavonoids and flavonoids. The major flavonoid pigments are derived from naringenin.

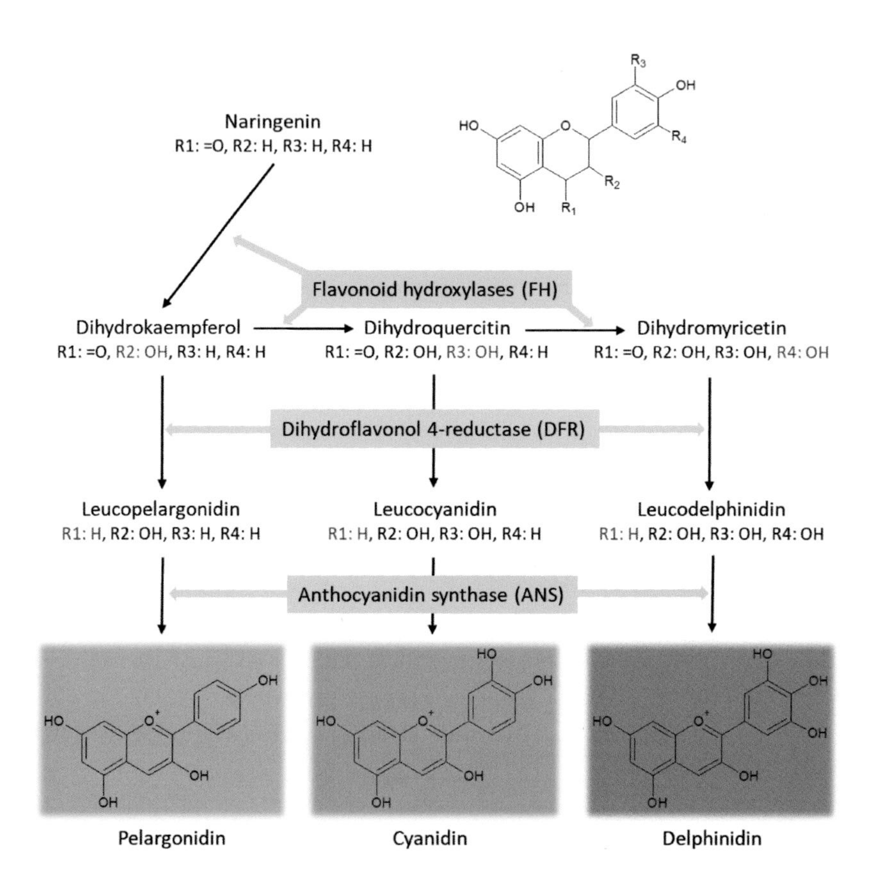

The biosynthetic grid leading from naringenin to anthocyanidins

A series of enzymic hydroxylations leads from naringenin to the flavonoid precursors of the vivid red, pink, mauve, violet, blue and purple anthocyanin pigments found in many petals, leaves, stems and fruits[200]. Dihydrokaempferol, dihydroquercitin and dihydromyricetin and their derivatives feed into a metabolic grid within which they are first converted, by the enzyme DFR[201], into the corresponding leucoanthocyanidins (leucopelargonidin, leucocyanidin and leucodelphinidin respectively). Leuco- here means these compounds are colourless, as are all the upstream precursors in anthocyanin biosynthesis. The next enzyme step carries out the critical conversion of the flavonol skeleton into the flavylium nucleus. The enzyme responsible, ANS, makes coloured anthocyanidins from leuco- precursors[202]. More than 30 anthocyanidin structures have been identified, the commonest of which are based on pelargonidin, cyanidin and delphinidin. Delphinidin is responsible for the blues and purplish colours of many flowers and fruits, including delphiniums, violets, grapes and berries. Cyanidin is found in many red berries, grapes, apples, plums, red cabbage and red onion. Pelargonidin is an orange pigment present in red geranium and other flowers, a range of fruits and kidney beans. Anthocyanidin colour and chemical stability are sensitive to pH. Delphinidin and cyanidin are red in acidic solution and blue at basic pH. Within plant cells, these compounds and the anthocyanins derived from them are generally located in the vacuole[203]. Leucoanthocyanidins and anthocyanidins are the precursors of catechins and condensed tannins, widely-distributed astringent compounds. Tannins deter pathogens and herbivores by complexing with macromolecules and inactivating enzymes. Tea, cocoa, chocolate, red wine and many fruits and vegetables are rich in catechins, which contribute to taste and nutritional effects.

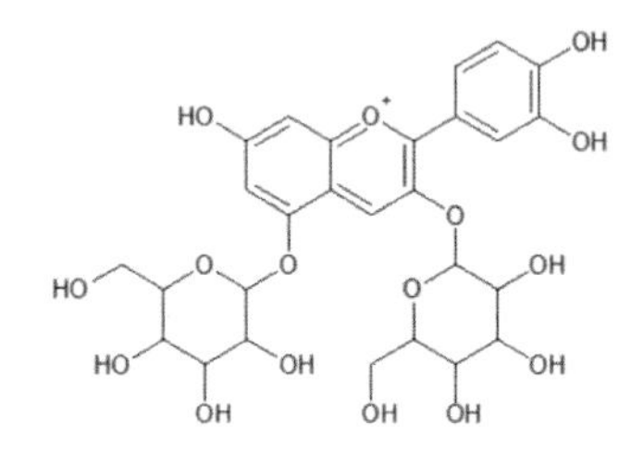

Cyanin (cyanidin 3,5-*O*-diglucoside)

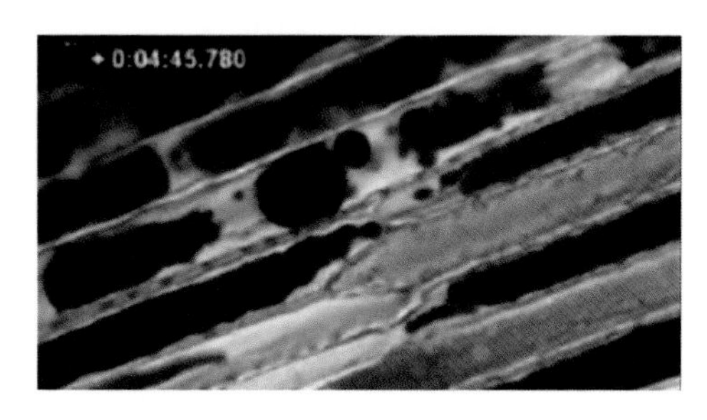

Anthocyanins in vacuoles of maize cells

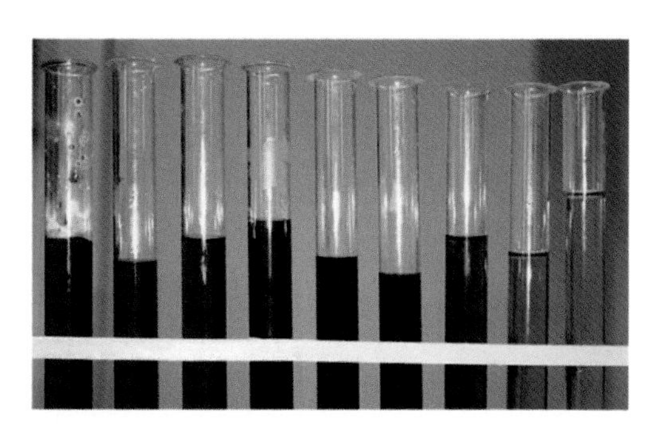

Acid pH Basic

Structure, vacuolar location and pH sensitivity of the common anthocyanin cyanin[204]

Continuing the task of disentangling the naming of flavonoids, anthocyanidins lose 'id' from the term as they are further modified to become anthocyanins. The earliest use of the word anthocyanin (1835) is attributed to the German pharmacist Ludwig Clamor Marquart[205]. The work of Arthur Ernest Everest[206] and others established that anthocyanins are members of the flavonoid family of plant secondary products, and that the fiery red of autumn leaves such as those of maples is due primarily to sugar conjugates (glucosides) of cyanidin. Your glass of red wine is an aqueous solution of cyanidin glucosides, mainly cyanin, and related anthocyanins (plus, of course, some other fun stuff like alcohol, fragrances, flavours and so on). The anthocyanin products of anthocyanidin conjugation with sugars may be subject to further biochemical modifications, including the addition of extra methyl, sugar and hydroxyl groups[207]. As well as structural variations, anthocyanin colours may be enhanced or altered by copigmentation with flavonols[208]. Some phenylpropanoids are not themselves coloured but act as optical brighteners to enhance the colours of other pigments – as happens, for example, in the glowing golden autumn foliage of the Ginkgo tree[209]. Combinations of phenylpropanoids with carotenoids or chlorophyll result in bronze and brown colours seen in some leaves and flowers and in the brown/purple fruits of transgenic tomato[210]. Anthocyanins and anthocyanidins, unlike carotenoids and chlorophylls, are generally quite water-soluble and this is reflected in where they live in the plant cell – not in the chloroplast, which is too hydrophobic, but in the wet environment of the vacuole[211]. Like anthocyanidins, the colours of anthocyanins are pH-sensitive, and changes in the generally acid environment of the cell vacuole can influence the perceived pigmentation of the tissue[212]. For example, activation of a hydrogen ion pump in the vacuole membrane causes an increase in vacuolar pH from 6.6 to 7.7 during opening of the 'Heavenly Blue'[213] variety of morning glory (*Ipomoea tricolor*) flowers, resulting in a change in the colour of petal anthocyanin from purplish red to blue[214]. In some cases, anthocyanins in vacuoles form aggregates with proteins and this has the effect of modifying and intensifying the colour of the tissue[215].

1. *Antirrhinum* petal pigmentation at (a) 25°C and (b) 15°C resulting from temperature-sensitive transposon excision; 2. Modification of red-flowered variety of petunia (a) by silencing the CHS gene (b-d); 3. The 'Blue' rose; 4. Inhibition of normal intense Heavenly Blue pigmentation of morning glory flower (a) by gene editing (b)

Just as the colour preferences of pollinating and seed-dispersing animals will have selected for the diversity of floral and fruit pigmentation during evolution, so too has selective breeding resulted in every conceivable shade of ornamental plant. *Antirrhinum* (snapdragon) is a model for genetic and molecular analysis of how modulations of phenylpropanoid metabolism and floral development lead to the diversity of colours, pigmentation patterning and petal morphology in ornamental flowers[216]. Gregor Mendel, Charles Darwin and William Bateson were among the giants of genetics and evolutionary biology to carry out inheritance studies on snapdragon. The foundations of the resources that established *Antirrhinum* as a favoured experimental subject were laid by the German botanist Erwin Bauer from 1907, and by 1966 there was a collection of over 500 mutants held at IPK, the plant culture institute in Gatersleben[217]. A particular feature of many of these mutants is their instability, which was subsequently shown to be the consequence of transposon insertion and excision, the system originally described for maize by McClintock. The *Antirrhinum* transposon *Tam1* was identified by its interaction with the gene *NIVEA*[218], which encodes the key enzyme of flavonoid biosynthesis CHS, and *Tam3* interacts with *PALLIDA*[219], the gene encoding the leucoanthocyanidin synthesis step in flower pigmentation. There are *Antirrhinum* lines in which temperature-sensitive transposition by *Tam3* results in white flowers at $25^{\circ}C$ and petal variegation at $15^{\circ}C$[220]. Detailed understanding of phenylpropanoid metabolism and its molecular regulation has opened the way to targetted modification of flower colour using the biotechnological tools of gene transfer, silencing, overexpression and editing[221]. One of the earliest such interventions involved reducing the activity of the gene for CHS in petunia[222]. Floriculturists have long dreamt of creating the blue rose and the blue chrysanthemum[223]. By genetically modifying the pathway of anthocyanidin synthesis, it has been possible to persuade rose petals to accumulate enough delphinidin, which they are normally unable to make, to turn blue(-ish)[224]. A current technological development which is causing much excitement in the biomedical and crop science communities is gene editing by the CRISPR/Cas 9 system. Among the recent reports of its successful application to alter floral colour development is strong suppression of blue pigmentation in morning glory as the result of editing the gene for DFR[225].

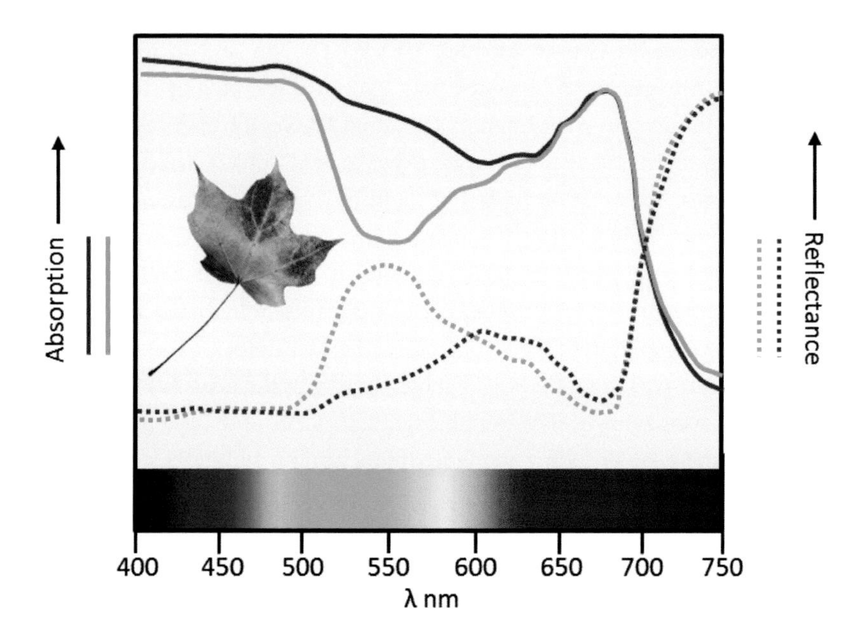

As anthocyanins accumulate in autumnal leaves of maple, reflectance of light at green wavelengths is reduced and orange-red reflectance increases. The pattern of light absorption (which drives photosynthesis) is essentially the reciprocal of reflectance[226]

Leaf anthocyanins are responsible for some of the most beautiful of the autumnal colours – reds, purples and even blues. As with floral pigmentation in ornamentals, the brilliant colours of autumn foliage in horticultural varieties of maple and other species have also been intensified by human selection. Young flush leaves in tropical forests are also often red or purple. Red is the colour of the lower leaf surfaces of many understory and floating plants, and of the thorns and spines of numerous species. Stress commonly causes foliage to blush. We may know a lot about the chemistry and genetics of anthocyanins, but the question as to why the vegetative parts of plants go red is difficult to answer[227]. Marco Archetti[228] has compiled as many as sixteen different hypotheses that address the question of the function of anthocyanins in the reddening response. These ideas can be divided into three main types. Plant scientists generally consider the role of anthocyanins is to act as sunblockers, shielding the tissue from the damaging effects of light at low temperatures, allowing a more efficient resorption of nutrients, especially nitrogen. Anthocyanins, and flavonoids in general, have antioxidant properties and will have a protective function in defending against stresses that cause a buildup of harmful ROS and free radicals[229]. Alternatively, or additionally, the entomologists and evolutionary biologists propose a coevolutionary explanation, in which red is a warning signal of the defensive status of the plant to animals, particularly feeding insects such as aphids. Whatever the explanation, humans can revel in the beauty of the colourful, if melancholy, fates of autumn leaves[230].

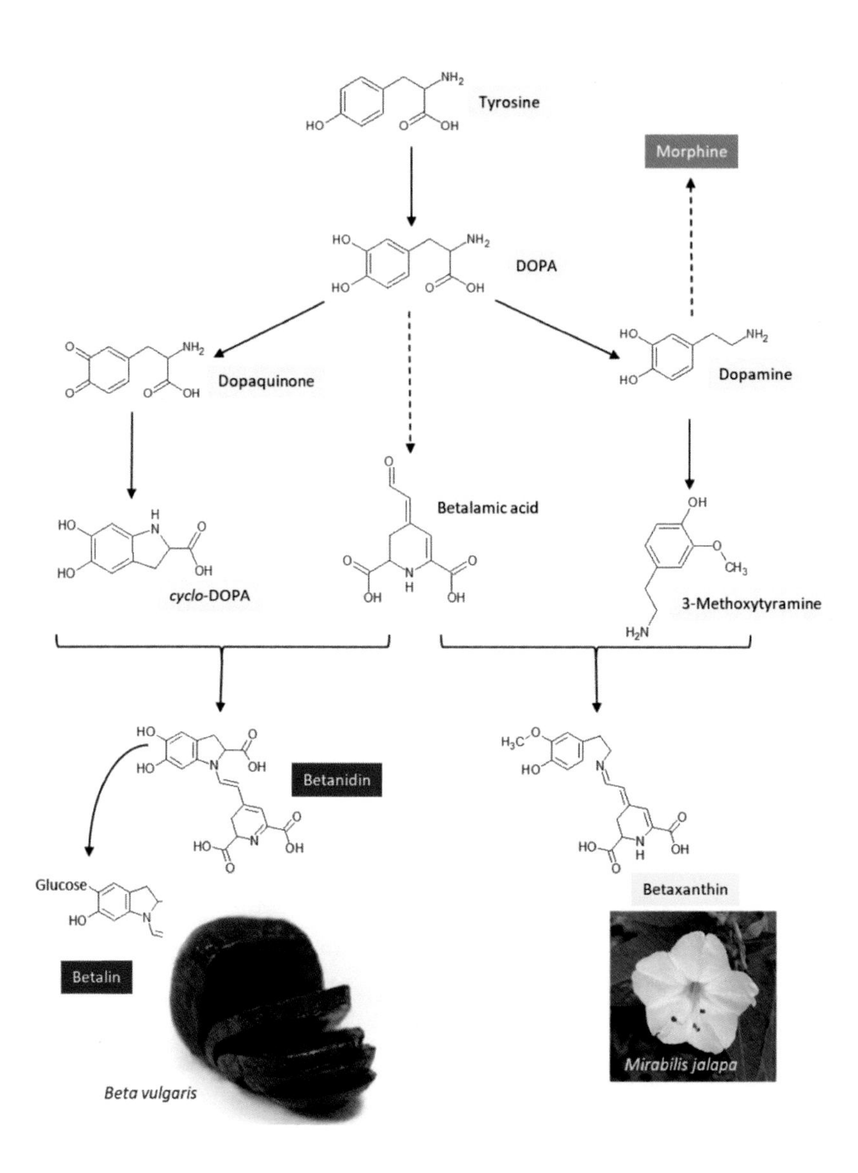

Biosynthesis of betalains from tyrosine[231]

Now here's a very curious little pigment fact. Flowers and other organs of species in families within the Caryophyllales (amaranths, cactuses and beets) owe their colours not to flavonoids or carotenoids, but to a chemically unrelated pigment family, the betalains, comprising the red betacyanins and the yellow betaxanthins. The intense colour of beetroots, for example, is due to betanidin, a glycosylated betalain. Yellow flowers of the four-o-clock plant, *Mirabilis jalapa*, are pigmented with betaxanthins. Betalains are nitrogenous water-soluble derivatives located in cell vacuoles, and they share the early stages of their biosynthesis with those leading to alkaloids such as morphine[232]. The curious fact is that betalains are absent from species that accumulate anthocyanins and vice versa[233]. Why this should be has long exercised plant systematists and biochemists, but recent research is beginning to shed some light on this mystery. It's agreed that production of floral anthocyanins, the biochemistry of which is highly conserved across the angiosperms, is the ancestral condition. Knowledge of betalain biosynthesis is incomplete but improving. The amino acid tyrosine is the source of betacyanins and betaxanthins and is converted to DOPA, probably by the enzyme tyrosinase. The betanidin precursor *cyclo*-DOPA is derived from DOPA, catalysed by a specific cytochrome P450. Betalamic acid, the common precursor of betacyanins and betaxanthins, is the product of a complex reaction catalysed by a dioxygenase enzyme, DODA. The evolutionary history of pigmentation in the Caryophyllales has been recreated by phylogenetic analysis of the cytochrome P450 and DODA gene families[234]. In this way it has been established that betalain-specific forms of cytochrome P450 and DODA initially appeared only once as the result of gene duplications at the evolutionary origin of betalain pigmentation, and that anthocyanin synthesis was suppressed, ultimately by gene elimination. Genes for betalain synthesis appear to have been subsequently lost from anthocyanic lineages within the Caryophyllales. There remain many unanswered questions about what evolutionary forces prompted the acquisition of, and reciprocal relationships between, betalain and anthocyanin pigmentation capacities, and drove the pattern of lineage radiation in these plant families.

Red sorghum[235]

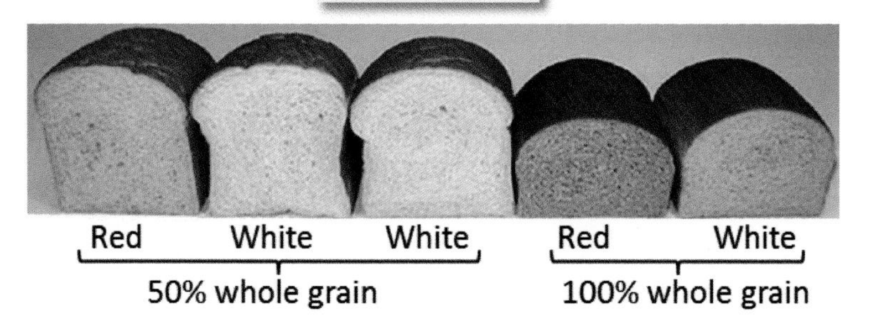

Loaves made with red and white wheat[236]

Earlier I discussed sorghum as a case-study in the relationship between pigment (in that case, the green of chlorophyll), agriculture and political history. Like maize and many other cereals, sorghum grains come in a range of colours from white to red to dark purple. This diversity of pigmentation is a consequence of genetic variation in phenylpropanoid biochemistry, and has a dramatic history which it is instructive to examine at length, as an episode in the tale of the three little pigments that has cultural, literary and socio-economic significance[237]. In *Red Sorghum* (1992), the celebrated novel by Nobel laureate Mo Yan [238], the crop (in this case grown for brewing) is at the centre of village community life ('It is the music of the universe, and it emanates from the red sorghum'). The vivid colour of the grain is a recurrent counterpoint to and — as Ning Yu has shown[239] — a synaesthetic metaphor for, the brutality and bloodshed experienced during the second Sino-Japanese war (1937-1945) and subsequent civil disturbances ('the angel of death, with lips as scarlet as sorghum and a smiling face the colour of golden corn'). There is a striking parallel here with other literary allusions to dark-coloured grain in the Western tradition. George Eliot uses it in *The Mill on the Floss*[240], another account of the life of a rural community founded on a cereal crop, that gets literally swept away in a human-mediated natural disaster[241]. The dark colouring of the miller's daughter, Maggie, which identifies her as a Tulliver rather than a Dodson – her father's daughter rather than her mother's – is explained in terms of the contrast between red and white varieties of wheat: 'the child's healthy enough', Mr Tulliver declares, responding to his daughter's unorthodox behaviour; 'there's nothing ails her. There's red wheat as well as white, for that matter, and some like the dark grain best.' Red wheat generally has a high content of bitter-tasting phenylpropanoids, and bread made from it has a darker, denser appearance and a strong flavour[242]. The long history of dark versus white bread is freighted with socio-political significance[243]. Having been traditionally cheaper than bread made from white wheat, many people – including Maggie, it would seem – acquired a taste for it. It's appropriate that the uneducated, undisciplined and unconventional Maggie is associated with red wheat, whereas her sibling, the more docile and socially-adept (but, perhaps, less engaging) Tom, is likened to the white.

Darnel (*Lolium temulentum*) mimicking red wheat[244]

There's a deep cultural, as well as biological, connection between seed pigmentation and bitter taste, reflecting a wider issue concerning the desirability of phytochemical richness in food and feed[245]. *Eating Bitterness* is the title of Michelle Loyalka's 2012 study of the 'pain, self-sacrifice, and fortitude' of the 200 million-strong floating population (*liudong*) of rural migrants that provide cheap labour for China's urban-based economic growth[246]. Jennifer McLaghan[247] has explored our relationship with bitter taste sensations in food (and how bitterness is under assault from the agrifood business in favour of blander, sweeter flavours). Western literature often plays on the incompatibility of sweet and bitter flavours and uses the contrast to distinguish between good and evil. In *The True Chronicle History of King Leir* (published 1605) – the principal source for Shakespeare's *King Lear* (completed 1606) – the eponymous monarch reflects on the contrast between the kind words and cruel actions of his elder daughters: 'Can … loue be reapt, where hatred hath bin sowne? ... Or Sugar grow in Wormwoods bitter stalke?/ It cannot be, they are too opposite'. For Joan of Arc in Shakespeare's *Henry VI Part 1* (1592), the cereal weed darnel is used to describe the bitter taste of a defeat brought about by treachery[248]. Using a clever arable metaphor entirely appropriate for someone of her rural upbringing, Joan, 'La Pucelle', likens the way darnel insinuates itself into the food chain to the way the French have smuggled themselves into the besieged city of Rouen, concealed in sacks of corn: 'Good morrow gallants. Want ye corn for bread?/ I think the Duke of Burgundy will fast/ Before he'll buy again at such a rate./ 'Twas full of darnel. Do you like the taste?' There is also, perhaps, the implication that some prefer beer and bread infused with the bitter 'taste' of darnel. Burgundy, whom the French regarded as a traitor for siding with the English, has been defeated by his own appetite.

Human chromosome 7

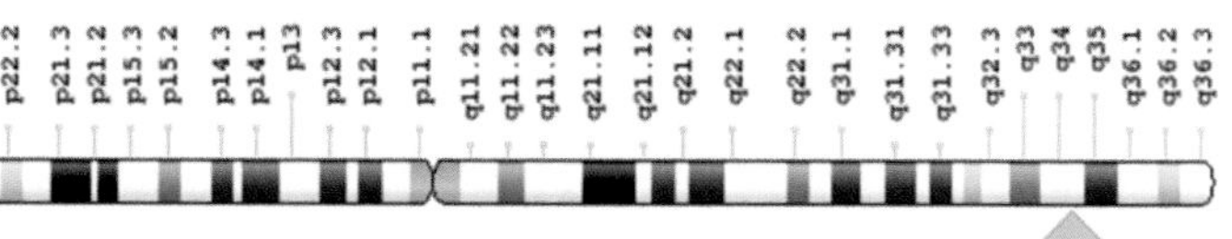

TAS2R38 gene

The gene for bitterness[249]

Taste evaluation broadly classifies sorghums into four groups, of increasing bitterness and astringency: white, red, black, tannin[250]. Taste preferences are complicated by intrinsic genetic differences between individuals at the TAS2R38 receptor locus in the human genome: the receptor is non-functional in some people, rendering them relatively insensitive to bitterness[251]. Which raises an intriguing but, alas, unanswerable question: do Maggie Tulliver and the Duke of Burgundy really prefer the bitterness of red wheat, or are they TAS2R38 null? The content of bitter phenylpropanoids (particularly free tannins) responsible for objectionable sensory attributes ranges widely across different red sorghum varieties[252]. Though it is often perceived to have adverse taste associations, the red-purple trait in cereal grains has agronomic benefits. Compounds in the metabolic pathway leading to dark pigmentation are potent inhibitors of fungal pathogens[253]. Because of its reduced susceptibility to pre-harvest sprouting, red wheat is better suited to wet and cold summers[254].

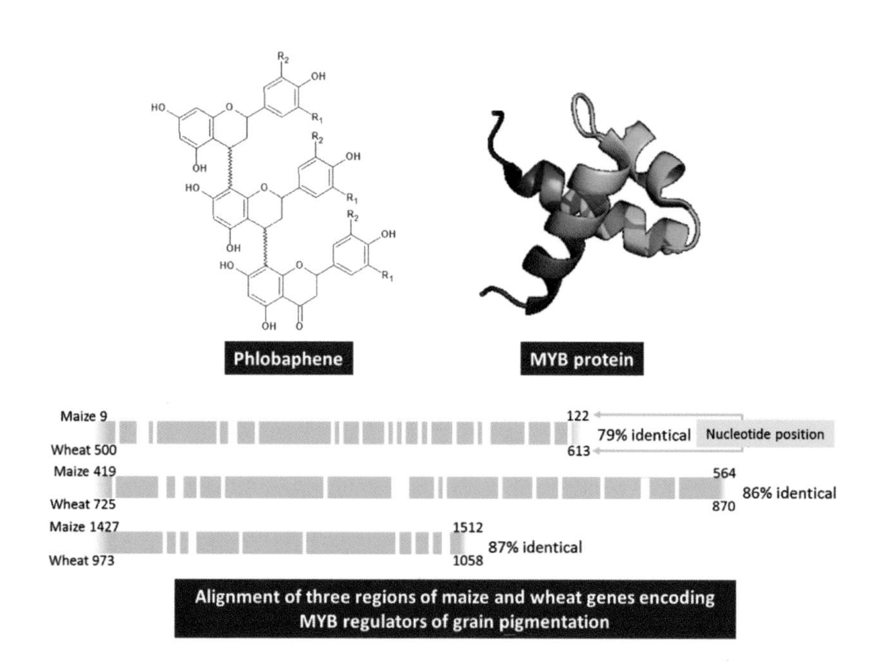

MYB regulator of phlobaphene synthesis in cereal grains and comparison of the maize and wheat genes that encode it[255]

The white testa character in cereals (not only wheat and sorghum, but also rice, maize and barley) is derived secondarily from the more primitive red-purple trait. The dark colour of the testa is due to phlobaphene, a reddish brown pigment derived from the bitter-tasting phenylpropanoid precursor naringenin[256]. Classical genetic analysis has shown the colour of sorghum testa to be under the control of two major genes, *R* and *Y*, with a few minor expression modifiers. The regulatory function of the sorghum *y1* (*yellowseed1*) locus was inferred from the absence of red colour from tissue with a non-functional *y1* allele[257]. The MYB transcription factor encoded by *y1* is functionally identical to the product of the orthologous maize gene *p1* (*pericarpcolor1*), which regulates accumulation of phlobaphene in cobs[258]. Himi and Noda[259] showed that the red grain colour gene in wheat also encodes a MYB-type transcription factor. So far as we know, the degree of sequence similarity between this wheat gene, *Tamyb10-A*, and the *MYB*s responsible for the red grain colour in sorghum and maize has not been reported. A BLAST alignment between maize *p1*, wheat *Tamyb10-A* and sorghum *y1 MYB* genes confirmed extremely close sequence matches[260], strongly supporting a common function. It's satisfying to discover that that *Red Sorghum*, *Henry VI Part 1* and *The Mill on the Floss*, three literary texts seemingly worlds apart, share a central thematic element not just through their respective texts but down to the deepest level of genes and molecules.

The purple compulsion

The mad desire for purple. With these words, Pliny the Elder[261] recognised the spell that the colour has cast from the earliest period of human history (perhaps as far back as 1750 BCE[262]). There's something about purple that invokes the most extreme behaviour in individuals and populations[263]. Clothe a Roman Emperor in Tyrian purple and, crazed with power, he 'bestride(s) the narrow world/ Like a Colossus'. At the other end of history, the era of modern dyes began in 1856 with Perkins's mauve, the aniline derivative that became all the rage in the fashionable world of mid-century London[264]. For much of history the 'mad lust' for purple could not be satisfied by shellfish of the genera *Murex* and *Thais* that yield the imperial pigment, nor by the chemistry that conjured dyestuffs from coal tar. Isaac Newton described the purplish colours at the end of the spectrum as indigo and violet[265]. In its day indigo dye, from leguminous plants of the genus *Indigofera,* and the Brassicaceous woad plant *Isatis,* was second in value only to Tyrian purple[266]. By an extraordinary twist of evolutionary fate, indigo and imperial purpurea are nitrogenous compounds that differ in molecular structure only by the latter's inclusion of two additional bromine atoms[267]. Lichens are another biological source of purplish dyes[268]. The blue-red acidity indicator litmus, familiar from school science labs, is an example. The traditional cloaks, variously described as red, mauve or blue, worn by the women of Fishguard in Wales as they repulsed an attempted French invasion in 1797 are said to have been dyed with the litmus-like lichen pigment orchil[269].

According to Friedrich Nietzsche[270] 'in individuals, insanity is rare; but in groups, parties, nations and epochs, it is the rule'. An extreme example of purple-fuelled irrationality was the tulip mania that engulfed the Netherlands in the 17th century. Tulip flowers owe their colours to anthocyanins, comprising rutosides of pelargonidin, delphinidin and cyanidin and their acetyl derivatives[271]. Charles MacKay's famous 1852 memoir[272] on popular delusions placed tulip mania firmly up there with the South Sea Bubble as an archetype of ludicrous herd behaviour. It forms the background to *The Black Tulip*, the novel by Alexandre Dumas (1850)[273], which captures the febrile atmosphere of the period through harrowing accounts of the lynching in 1672 of the de Witt brothers by a baying mob, the obsessive quest of Cornelius van Baerle to win the prize for breeding the flower of the novel's title, and the malicious madness of his jealous rival Isaac Boxtel. Dumas gives a rather clear description of the plant breeding strategy that would lead to what the tulip world considered to be 'as chimerical as the black swan of Horace or the white raven of French tradition'. Life ultimately imitates art: after years of crossing and selecting progressively darker and darker purple lines, the Dutch grower Geert Hageman produced the variety 'Paul Scherer' in 1997, the nearest thing to a black tulip yet[274].

In 1759 Samuel Johnson wrote 'The business of a poet is to examine, not the individual, but the species; to remark general properties and large appearances: he does not number the streaks of the tulip'[275]. It takes a brave person to disagree with the good doctor; but in this case, I venture he is mistaken and has overlooked the deep insights into human obsessions and follies to be gained by considering tulip flowers and 'the plum history of the color purple'[276].

NOTES AND SOURCES

[178] Rainbow corn from https://pixabay.com/en/corn-harvest-colors-2798026/ reproduced under CC0 Creative Commons [accessed 25 June 2018].
[179] I'm not one, by the way.
[180] Keller EF. 1983. *A Feeling for the Organism.* New York: Freeman
[181] Gardner M. 1950. The hermit scientist. Antioch Review 10: 447-457.
[182] Comfort NC. 1999. 'The real point is control': the reception of Barbara McClintock's controlling elements. Journal of the History of Biology 32: 133–62.
[183] Jones RN. 2005. McClintock's controlling elements: the full story. Cytogenetic and Genome Research 109: 90–103.
[184] One of the plant products closest in chemical structure to phenylpropane is anethole. It's a major component of the flavours and fragrances of essential oils from a number of species, including anise, fennel, liquorice and camphor. Anethole is more soluble in alcohol than in water, and this accounts for the formation of the opaque milky emulsion on dilution of anise-flavoured spirits such as ouzo, rakı, sambuca and absinthe.

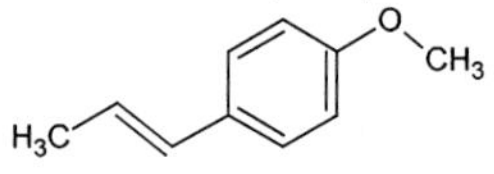

Anethole

[185] Jones RL, Ougham H, Thomas H, Waaland S. 2012. *The Molecular Life of Plants.* Chichester: Wiley.
[186] The etymological origins of words derived from the element 'flav-' are curious (https://www.etymonline.com/ [accessed 18 April 2018]). It comes from *flavus*, Latin for yellow, but shares a deep source with the word 'blue' in the Proto-Indo-European roots *bhel-* or *ghel-*, meaning 'light-coloured, blue, blond, yellow, shining, flashing, burning'. Several languages and cultures categorise yellow and blue together, but why this should be is anyone's guess (Ball P. 2001. *Bright Earth: the Invention of Colour.* London: Vantage). Many flavonoids are indeed yellow, but many others, notably the anthocyanins, are blue, red, purple, brown and colours in between.
[187] Plants protect themselves against herbivores, microbial pathogens and competing neighbours by accumulating a range of miscellaneous phenylpropanoid-related products which include stilbenes and coumarins. Stilbenes, which lack the heterocyclic ring of flavonoids, are synthesised by plants in response to damage. Grapes, pines and legumes are a particularly rich source of the resveratrol family of stilbenes, which function as pathogen defence agents (phytoalexins) and have cardioprotective and anticarcinogenic properties.

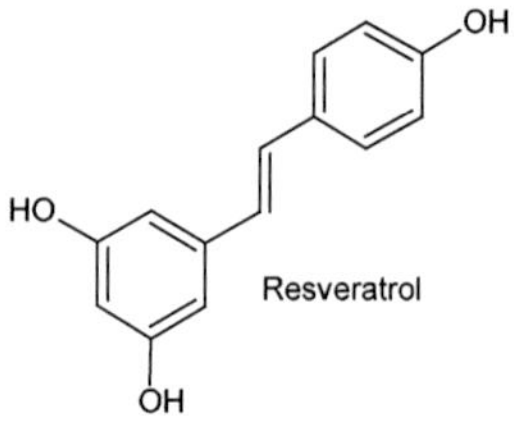

Resveratrol

Coumarins, phenylpropanoids from which one of the flavan phenyls is absent, are fragrant, bitter appetite-suppressers. The anticoagulant warfarin (coumadin), which is used as a rat poison and blood thinner, is a chemically-modified coumarin derivative.
[188] Isoflavones differ from flavones in the point of attachment between the isocyclic and phenyl ring. Phytoestrogens are much-studied biologically active isoflavonoids with affinities for the oestrogen hormone receptors of animals, and have been used as medical treatments and dietary supplements (Dixon RA, Pasinetti GM. 2010. Flavonoids and isoflavonoids: from plant biology to agriculture and neuroscience. Plant Physiology 154: 453-457).
[189] Ono E, Fukuchi-Mizutani M, Nakamura N, Fukui Y, Yonekura-Sakakibara K, Yamaguchi M, Nakayama T, Tanaka T, Kusumi T, Tanaka Y. 2006. Yellow flowers generated by expression of the aurone biosynthetic pathway. Proceedings of the National Academy of Sciences 103: 11075-11080.
[190] Grotewold E. (ed) 2006. *The Science of Flavonoids*. NY: Springer.
[191] Bruckner V, Szent-Györgyi A. 1936. Chemical nature of citrin. Nature 138: 1057.
[192] Pauling L. 1939. Recent work on the configuration and electronic structure of molecules. Fortschritte der Chemie Organischer Naturstoffe 3: 203–235.
[193] Tohge T, Watanabe M, Hoefgen R, Fernie AR. 2013. The evolution of phenylpropanoid metabolism in the green lineage. Critical Reviews in Biochemistry and Molecular Biology 48: 123-152.
[194] Weng JK, Chapple C. 2010. The origin and evolution of lignin biosynthesis. New Phytologist 187: 273-285.
[195] A useful database of flavonoids is http://metabolomics.jp/wiki/Category:FL [accessed 28 April 2016].
[196] Here are KEGG metabolic maps for flavonoid biosynthesis from phenylalanine: http://www.genome.jp/kegg-bin/show_pathway?map00940 and http://www.genome.jp/kegg-bin/show_pathway?map00941 [accessed 20 April 2018].
[197] These amino acids are classified as aromatic because their structure includes a six-membered ring with a conjugated bond system (and not because they are particularly fragrant).
[198] MacDonald MJ, D'Cunha GB. 2007. A modern view of phenylalanine ammonia lyase. Biochemistry and Cell Biology 85: 273-282.
[199] PAL from monocots also has TAL activity, using tyrosine as a substrate to produce 4-coumaric acid (Rosler J, Krekel F, Amrhein N, Schmid J. 1997. Maize phenylalanine ammonia-lyase has tyrosine ammonia-lyase activity. Plant Physiology 113: 175-179).
[200] Here's the KEGG map for anthocyanin biosynthesis: http://www.genome.jp/kegg-bin/show_pathway?map00942 [accessed 20 April 2018].
[201] Heller W, Forkmann G, Britsch L, Grisebach H. 1985. Enzymatic reduction of (+)-dihydroflavonols to flavan-3,4-cis-diols with flower extracts from *Matthiola incana* and its role in anthocyanin biosynthesis. Planta 165: 284-287.
[202] Saito K, Kobayashi M, Gong Z, Tanaka Y, Yamazaki M. 1999. Direct evidence for anthocyanidin synthase as a 2-oxoglutarate-dependent oxygenase: molecular cloning and functional expression of cDNA from a red forma of *Perilla frutescens*. Plant. Journal 17: 181-189.

[203] Pourcel L, Irani NG, Lu Y, Riedl K, Schwartz S, Grotewold E. 2010. The formation of anthocyanic vacuolar inclusions in *Arabidopsis thaliana* and implications for the sequestration of anthocyanin pigments. Molecular Plant 3: 78-90.

[204] Anthocyanin-containing vacuoles in a screen shot from a time-lapse video of maize lemma cells (Irani NG, Grotewold E. 2005. Light-induced morphological alteration in anthocyanin-accumulating vacuoles of maize cells. BMC Plant Biology 5(1):7; video from Wikimedia, http://tinyurl.com/y89mm2w4 [accessed 12 May 2018]).

[205] Everest AE. 1915. The anthocyan pigments. Science Progress in the Twentieth Century (1906-1916) 9: 597-612.

[206] Everest AE. 1914. The production of anthocyanins and anthocyanidins. Proceedings of the Royal Society of London. Series B. 87: 444-452.

[207] Castaneda-Ovando A, Pacheco-Hernández M de L, Páez-Hernández ME, Rodríguez JA, Galán-Vidal CA. 2009. Chemical studies of anthocyanins: a review. Food Chemistry 113: 859-871. Glycosylation and other modifications to the flavonoid skeleton occur widely throughout phenylpropanoid metabolism.

[208] Boulton, R. 2001. The copigmentation of anthocyanins and its role in the colour of red wine: a critical review. American Journal of Enology and Viticulture 52: 67-87.

[209] Matile P. 2000. Biochemistry of Indian summer: physiology of autumnal leaf coloration. Experimental Gerontology 35: 145-158.

[210] Robinson GM, Robinson R. 1932. A survey of anthocyanins. II. Biochemical Journal 26: 1647-1664; Gonzali S, Mazzucato A, Perata P. 2009. Purple as a tomato: towards high anthocyanin tomatoes. Trends in Plant Science 14: 237-241.

[211] Tanaka Y, Sasaki N, Ohmiya A. 2008. Biosynthesis of plant pigments: anthocyanins, betalains and carotenoids. Plant Journal 54: 733-749.

[212] Richard Willstätter pointed out that the same pigment, cyanidin 3,5-*O*-diglucoside (cyanin), is responsible for the blue of cornflowers and the red of roses. He attributed the variety of flower colours to differences in cell pH environments. (Gould K, Davies KM, Winefield C. 2008. *Anthocyanins: Biosynthesis, Functions, and Applications*. NY: Springer). We now know that self-association, formation of complexes with metal ions, copigmentation with other phenylpropanoids, and intramolecular sandwich-type stacking also contribute to colour variation and stabilisation of anthocyanins in aqueous solution.

[213] The 'Heavenly Blue' complex from the petals of *Ipomoea* is an example of how extensively the anthocyanin skeleton may be decorated. The molecular basis of the intense blue colour is a combination of peonidin with six molecules of glucose and three molecules of caffeic acid (Kondo T, Kawai T, Tamura H, Goto T. 1987. Structure determination of heavenly blue anthocyanin, a complex monomeric anthocyanin from the morning glory *Ipomoea tricolor*, by means of the negative NOE method. Tetrahedron Letters 28: 2273-2276). The pigment of bluebells and Asiatic dayflowers is malonylawobanin, built from delphinidin diglucoside complexed with one molecule of malonic acid and one of coumaric acid (Goto T, Kondo T, Tamura H, Takase S. 1983. Structure of malonylawobanin, the real anthocyanin present in blue-colored flower petals of *Commelina communis*. Tetrahedron Letters 24: 4863-4836).

[214] Yoshida K, Kawachi M, Mori M, Maeshima M, Kondo M, Nishimura M, Kondo T. 2005. The involvement of tonoplast proton pumps and Na^+ (K^+)/H^+ exchangers in the change of petal color during flower opening of morning glory, *Ipomoea tricolor*. Plant and Cell Physiology 46: 407-415.

[215] Zhang H, Wang L, Deroles S, Bennett R, Davies K. 2006. New insight into the structures and formation of anthocyanic vacuolar inclusions in flower petals. BMC Plant Biology 6: 29; Okamura M, Nakayama M, Umemoto N, Cano EA, Hase Y, Nishizaki Y, Sasaki N, Ozeki Y. 2013. Crossbreeding of a metallic color carnation and diversification of the peculiar coloration by ion-beam irradiation. Euphytica 191: 45-56.
[216] Schwarz-Sommer Z, Davies B, Hudson A. 2003. An everlasting pioneer: the story of *Antirrhinum* research. Nature Reviews Genetics 4: 655-664.
[217] Stubbe H. 1966. *Genetik und Zytologie von* Antirrhinum *L. sect.* Antirrhinum. Jena: Gustav Fischer.
[218] Bonas U, Sommer H, Saedler H. 1984. The 17-Kb Tam-1 element of *Antirrhinum majus* induces a 3-bp duplication upon integration into the chalcone synthase gene. EMBO Journal 3: 1015–1019.
[219] Martin C, Carpenter R, Sommer H, Saedler H, Coen ES. 1985. Molecular analysis of instability in flower pigmentation of *Antirrhinum majus*, following isolation of the *Pallida* locus by transposon tagging. EMBO Journal 4: 1625–1630.
[220] Uchiyama T, Hiura S, Ebinuma I, Senda M, Mikami T, Martin C, Kishima Y. 2013. A pair of transposons coordinately suppresses gene expression, independent of pathways mediated by siRNA in *Antirrhinum*. New Phytologist 197: 431-440.
[221] Darqui FS, Radonic LM, Hopp HE, Bilbao ML. 2017. Biotechnological improvement of ornamental plants. Ornamental Horticulture 23: 279-288; Kishi-Kaboshi M, Aida R, Sasaki K. 2018. Genome engineering in ornamental plants: Current status and future prospects. Plant Physiology and Biochemistry https://doi.org/10.1016/j.plaphy.2018.03.015.
[222] Van der Krol AR, Lenting PE, Veenstra J, van der Meer IM, Koes RE, Gerats AG, Mol JN, Stuitje AR. 1988. An anti-sense chalcone synthase gene in transgenic plants inhibits flower pigmentation. Nature 333: 866.
[223] Yoshida K, Mori M, Kondo T. 2009. Blue flower color development by anthocyanins: from chemical structure to cell physiology. Natural Product Reports 26: 884–915.
[224] Katsumoto Y, Fukuchi-Mizutani M, Fukui Y, Brugliera F, Holton TA, Karan M, Nakamura N, Yonekura-Sakakibara K, Togami J, Pigeaire A, Tao GQ. 2007. Engineering of the rose flavonoid biosynthetic pathway successfully generated blue-hued flowers accumulating delphinidin. Plant and Cell Physiology 48: 1589-1600.
[225] Watanabe K, Kobayashi A, Endo M, Sage-Ono K, Toki S, Ono M. 2017. CRISPR/Cas9-mediated mutagenesis of the dihydroflavonol-4-reductase-B (DFR-B) locus in the Japanese morning glory *Ipomoea* (*Pharbitis*) *nil*. Scientific Reports. 7: 10028.
[226] Based on data from Gitelson AA, Merzlyak MN, Chivkunova OB. 2001. Optical properties and nondestructive estimation of anthocyanin content in plant leaves. Photochemistry and Photobiology 74: 38-45.
[227] Archetti M, Döring TF, Hagen SB, Hughes NM, Leather SR, Lee DW, Lev-Yadun S, Manetas Y, Ougham HJ, Schaberg PG, Thomas H. 2009. Unravelling the evolution of autumn colours – an interdisciplinary approach. Trends in Ecology and Evolution 24: 166-173.
[228] Archetti M. 2009. Classification of hypotheses on the evolution of autumn colours. Oikos 118: 328-333.
[229] Plant products rich in beneficial antioxidant phenylpropanoids have long been of medical, nutritional (and cosmetic) interest: Zhang H, Tsao R. 2016. Dietary polyphenols,

oxidative stress and antioxidant and anti-inflammatory effects. Current Opinion in Food Science 8: 33-42.
[230] Griffith C. 2004. *Fall*. New York: Powerhouse Books.
[231] Public domain (CC0) image of sliced beetroot from https://toppng.com/photo/11890/beetroot-sliced [accessed 25 June 2018]; *Mirabilis jalapa* from https://commons.wikimedia.org/wiki/File:Mirabilis_japonica_(28735952023).jpg reproduced under Creative Commons CC BY-SA [accessed 25 June 2018].
[232] Grotewold E. 2006. The genetics and biochemistry of floral pigments. Annual Review of Plant Biology 57: 761-780.
[233] Stafford HA. 1994. Anthocyanins and betalains: evolution of the mutually exclusive pathways. Plant Science 101: 91–98.
[234] Brockington SF, Yang Y, Gandia-Herrero F, Covshoff S, Hibberd JM, Sage RF, Wong GKS, Moore MJ, Smith SA. 2015. Lineage-specific gene radiations underlie the evolution of novel betalain pigmentation in Caryophyllales. New Phytologist 207: 1170–1180.
[235] Oil painting by Wang Jian and Yin Luwen (2014; http://tinyurl.com/y93dsguw [accessed 27 April 2018]).
[236] Image from Ransom JK, Berzonsky WA, Sorenson BK. 2015. *Hard white wheat*. Publication A-1310 of NDSU Extension Service (https://www.ag.ndsu.edu/pubs/plantsci/smgrains/a1310.pdf [accessed 25 June 2018]), reproduced under Creative Commons CC BY-NC-SA with permission (thanks to Becky Koch and Deborah Tanner, NDSU).
[237] I'm grateful to Jayne Archer for insights into literary allusions to grain colour and bitterness.
[238] Yan, M. 1992. *Red Sorghum* (translated by H Goldblatt). London: Arrow Books.
[239] Yu N. 2003. Synesthetic metaphor: a cognitive perspective. Journal of Literary Semantics 32: 19-34.
[240] Eliot G. 1860. *The Mill on the Floss*. London: Vintage.
[241] Archer JE, Marggraf Turley R, Thomas H. 2015. 'Moving accidents by flood and field': the arable and tidal worlds of George Eliot's *The Mill on the Floss*. English Literary History 82: 701-28.
[242] Challacombe CA, Abdel-Aal EM, KousSeetharamana K, Duizer LM. 2012. Influence of phenolic acid content on sensory perception of bread and crackers made from red or white wheat. Journal of Cereal Science 56: 181-188.
[243] Wilson B. 2008. *Swindled: From Poison Sweets to Counterfeit Coffee - The Dark History of the Food Cheats*. London: John Murray; Collingham L. 2017. *The Hungry Empire: How Britain's Quest for Food Shaped the Modern World*. London: Bodley Head.
[244] Image of *L. temulentum* from Natura Italiana (web site: http://tinyurl.com/pbgvdy5 [accessed 27 April 2018]), reproduced by kind permission of Professor Franco Caldararo.
[245] Provenza FD, Meuret M, Gregorini P. 2015. Our landscapes, our livestock, ourselves: Restoring broken linkages among plants, herbivores, and humans with diets that nourish and satiate. Appetite 95: 500-519.
[246] Loyalka M. 2012. *Eating Bitterness: Stories from the Front Lines of China's Great Urban Migration*. Berkeley and Los Angeles: University of California Press.
[247] McLagan J. 2014. *Bitter*. HarperCollins Canada.
[248] Thomas H, Archer JE, Marggraf Turley R. 2016. Remembering darnel, a forgotten plant of literary, religious and evolutionary significance. Journal of Ethnobiology 36: 29-44.

[249] Disgust from https://commons.wikimedia.org/wiki/File:Disgust_expression_cropped.jpg (Creative Commons CC BY [accessed 25 June 2016]). Chromosomal location of TAS2R38 gene from https://ghr.nlm.nih.gov/gene/TAS2R38#location [accessed 25 June 2018].
[250] Dykes L, Rooney LW. 2007. Phenolic compounds in cereal grains and their health benefits. Cereal Foods World 52: 105-111; Kobue-Lekalake RI, Taylor JRN, de Kock HL. 2007. Effects of phenolics in sorghum grain on its bitterness, astringency and other sensory properties. Journal of the Science of Food and Agriculture 87: 1940–1948.
[251] Dinehart ME, Hayes JE, Bartoshuk LM, Lanier SL, Duffy VB. 2006. Bitter taste markers explain variability in vegetable sweetness, bitterness, and intake. Physiology and Behavior 87: 304–13.
[252] Kobue-Lekalake et al. (2007).
[253] Bate-Smith EC. 1969. Luteoforol (3',4,4',5,7-pentahydroxyflavan) in *Sorghum vulgare* L. Phytochemistry 8: 1803–1810; Snyder BA, Nicholson RL. 1990. Synthesis of phytoalexins in sorghum as a site-specific response to fungal ingress. Science 248: 1637–1639.
[254] Groos C, Gay G, Perretant M, Gervais L, Bernard M, Dedryver F, Charmet G. 2002. Study of the relationship between pre-harvest sprouting and grain color by quantitative trait loci analysis in a white×red grain bread-wheat cross. Theoretical and Applied Genetics 104: 39–47.
[255] Generic model of plant-type MYB protein from https://en.wikipedia.org/wiki/MYB_(gene)#/media/File:Protein_MYB_PDB_1guu.png (CC BY-SA 3.0 [accessed 25 June 2018]). DNA sequence alignments by NCBI BLAST (https://blast.ncbi.nlm.nih.gov/Blast.cgi [accessed 25 June 2018].
[256] Miyamoto T, Everson EH. 1958. Biochemical and physiological studies of wheat seed pigmentation. Agronomy Journal 50: 733–734; Himi E, Noda K. 2005. Red grain colour gene (*R*) of wheat is a Myb-type transcription factor. Euphytica 143: 239–42; Boddu J, Svabek C, Ibraheem F, Jones AD, Chopra S. 2005. Characterization of a deletion allele of a sorghum *Myb* gene *yellow seed1* showing loss of 3-deoxyflavonoids. Plant Science 169: 542–552. Naringenin and its derivatives give grapefruit its bitter taste (Esaki S, Nishiyama K, Sugiyama N, Nakajima R, Takao Y, Kamiya S. 1994. Preparation and taste of certain glycosides of flavanones and of dihydrochalcones. Bioscience Biotechnology and Biochemistry 58: 1479–1485). Like many flavonoids, naringenin has bioactive properties that have been extensively studied for potential medical applications (Rao V, Kiran SD, Rohini P, Bhagyasree P. 2017. Flavonoid: a review on naringenin. Journal of Pharmacognosy and Phytochemistry 6: 2778-2783).
[257] Boddu et al. (2005).
[258] Ibraheem F, Gaffoor I, Tan Q, Shyu C, Chopra S. 2015. A sorghum MYB transcription factor induces 3-deoxyanthocyanidins and enhances resistance against leaf blights in maize. Molecules 20: 2388-2404.
[259] Himi and Noda (2005).
[260] Helen Ougham carried out the sequence alignments - thanks, as ever.
[261] Pliny the Elder. 1855. *Natural History* IX xxxvi 126. tr. John Bostock. London. Taylor and Francis http://tinyurl.com/ycnhp7ey [accessed 2 April 2018].
[262] Elliott C. 2008. Purple pasts: color codification in the Ancient World. Journal of the American Bar Foundation 33: 173-194.

[263] And their deities: Shug Avery in Alice Walker's *The Color Purple* (1982; NY: Harcourt Brace Jovanovich) declared that it displeases God 'if you walk by the color purple in a field somewhere and don't notice it'.

[264] Ball (2008).

[265] Purple was an anomaly in Newton's colour theory. He knew that it could be made by combining red and blue paints; but red and blue are at the opposite ends of the spectrum. There's a continuing debate about why he introduced indigo into the sequence of rainbow colours. Biernson (1972) suggested that what Newton called indigo would be called violet in modern colour terminology, and his violet would be our purple (Biernson G. 1972. Why did Newton see indigo in the spectrum? American Journal of Physics 40: 526-33). A more widely held opinion is that Newton, an alchemical mystic by inclination, added two new terms - orange and indigo - to the traditional five of the medieval rainbow so that the colours would be 'divided after the manner of a Musical Chord' (McLaren K. 1985. Newton's indigo. Color Research and Application 10: 225-229). Newton adopted a pragmatic solution to the problem by joining the red and blue ends of the linear spectrum, through purple as an intermediate, to form a colour wheel. He was far from the first to represent colours in this way - colour wheels have been identified as far back as the 13th century, and the history of colour theory is replete with versions of similar circular spectra (Schmid F. 1948. The color circles by Moses Harris. The Art Bulletin 30: 227-230).

[266] Finlay V. 2002. *Colour: Travels through the Paintbox*. London: Hodder and Stoughton.

[267] Głowacki ED, Voss G, Leonat L, Irimia-Vladu M, Bauer S, Sariciftci NS. 2012. Indigo and Tyrian purple–from ancient natural dyes to modern organic semiconductors. Israel Journal of Chemistry 52: 540-551.

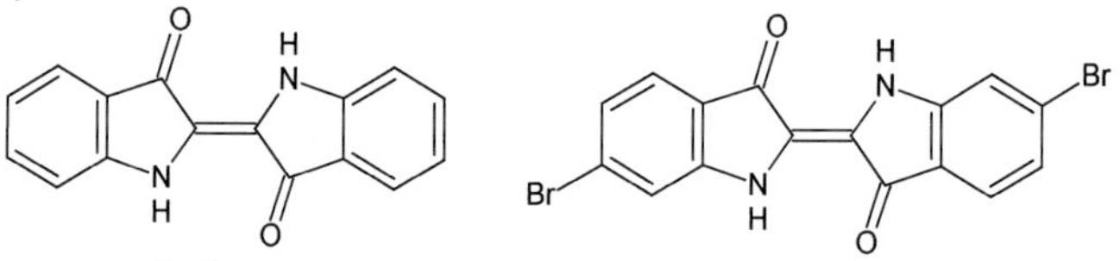

Indigo Tyrian purple

[268] Beecken H, Gottschalk EM, v Gizycki U, Krämer H, Maassen D, Matthies HG, Musso H, Rathjen C, Zdhorszky UI. 2003. Orcein and litmus. Biotechnic and Histochemistry 78: 289-302; Ferreira ES, Hulme AN, McNab H, Quye A. 2004. The natural constituents of historical textile dyes. Chemical Society Reviews 33: 329-336.

[269] St Clair K. 2016. *The Secret Lives of Colour*. London: Murray.

[270] Nietzsche F. 1886. *Beyond Good and Evil* (1998 edition). Oxford: World's Classics.

[271] Nakayama M, Okada M, Taya-Kizu M, Urashima O, Kan Y, Fukui Y, Koshioka M. 2004. Coloration and anthocyanin profile in tulip flowers. Japan Agricultural Research Quarterly: JARQ 38: 185-190. Rutoside (rutin) is a flavonoid diglucoside derived from naringenin.

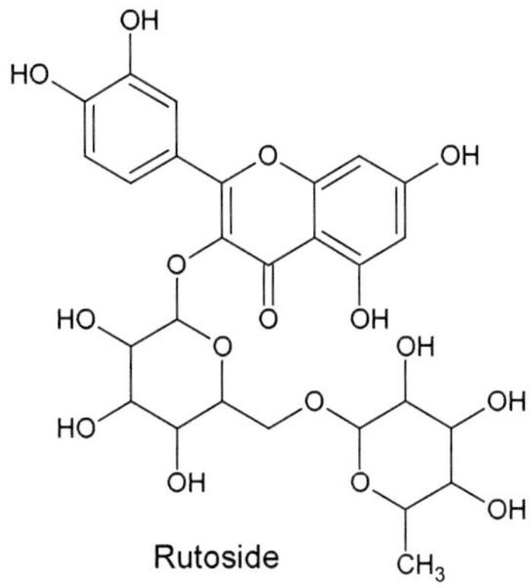

Rutoside

[272] Mackay C. 1852. *Memoirs of Extraordinary Popular Delusions and the Madness of Crowds*. London: National Illustrated Library.
[273] Dumas A. 1850. *The Black Tulip* (2003 edition) London: Penguin.
[274] As well as the Black Tulip, there are many purportedly black flower varieties: Black Narcissus, Black Iris, Black Petunia and so on. The garden designer and writer Karen Platt has published extensively on black plants. *Black Magic and Purple Passion* (3rd edition 2004), describes 2750 dark garden plants. Numerous fruits and seeds are identified as black: blackberry, blackcurrant, black gram, black cherry for example. A bible-black bloom or a forbidden crow-black fruit radiates an erotic frisson. But none of these blacks is truly black - they are all expressions of hyperaccumulated darkly intense purple, red and brown flavonoids.
[275] Samuel Johnson,1759. *The History of Rasselas, Prince of Abissinia*. London: Dodsley, Dodsley and Johnston; Folkenflik R. 1978. The tulip and its streaks: contexts of Rasselas X. ARIEL: A Review of International English Literature. Apr 1; 9(2). Johnson seemed to have a thing about tulips. Does anyone know what he meant by this? 'A tiger newly imprisoned is indeed more formidable, but not more angry, than Jack Tulip, withheld from a florist's feast' (Essay 48. The bustle of idleness described and ridiculed, posted in The Idler 1759).
[276] Elliott (2008).

5. FADEOUT

Concerning black, grey, brown and no colour at all[277]

Vivid black, take me back to the source

Maggie Nicols (*Vivid black*)[278]

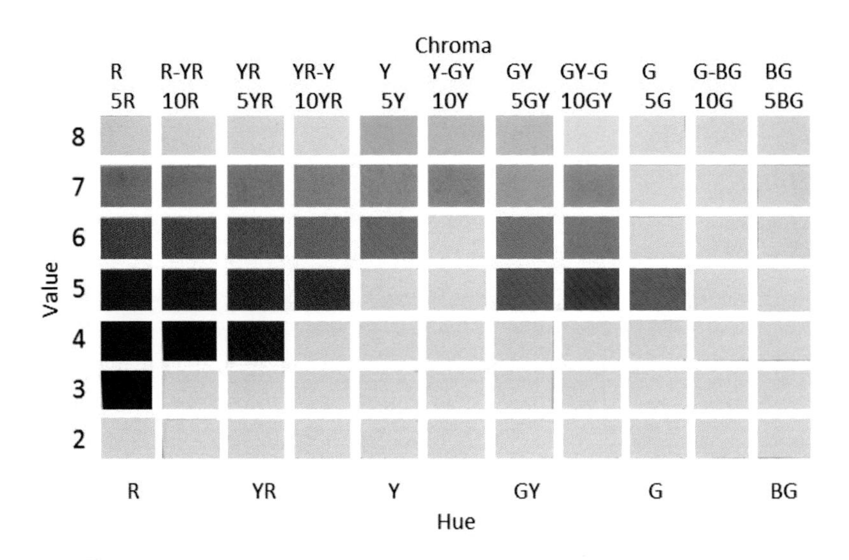

Red-Yellow-Green section of the Munsell System[279] covering the colour space occupied by most plant pigments

Are black and white colours? If so, how do they fit with the one-dimensional Newtonian spectrum and the two-dimensional colour wheel? Throughout history, from the Greeks to the Byzantines, from Goethe to Matisse[280], alternatives to Newton have ordered colours with reference to lightness and affinity with white (yellow, red, orange) through to darkness and proximity to black (blue, purple). Colour theory reconciles these different models by adding a third dimension, representing *value*, the quality of darkness (with black as a value of 0) to lightness (white with a value of 10) of a particular intensity (*chroma*) of a particular colour (*hue*). Such models (of which there have been many variants[281]) allow colours like brown, pink and grey, which otherwise lie outside the primary spectrum, to be located in colour space[282]. What does this mean for the Tale of the Three Little Pigments? It invites us to consider what happens to the pigments of a tissue or organ whose colour moves along the value dimension, either developmentally or pathologically, towards black or white as the final destination. It takes in, inter alia, the question as to what it is that organisms such as pollinating insects and birds perceive when the pigmentation of the plant structures they interact with is visible to them but outside the colour range to which the human eye is sensitive. It also raises the question of biological colour without pigments. And underlying all is time: in the end, biology returns to darkness or light[283].

Eumelanin monomer

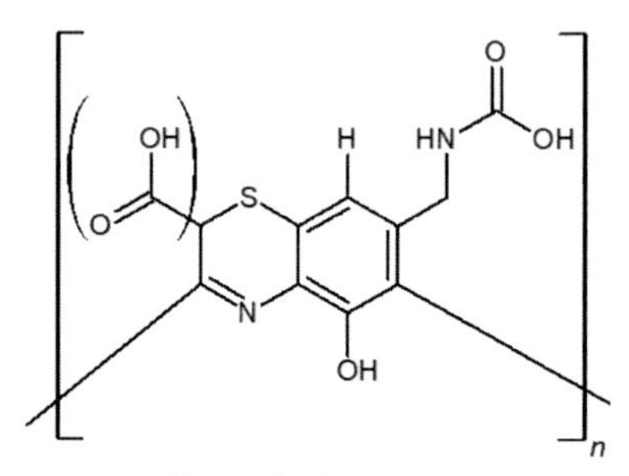

Pheomelanin monomer

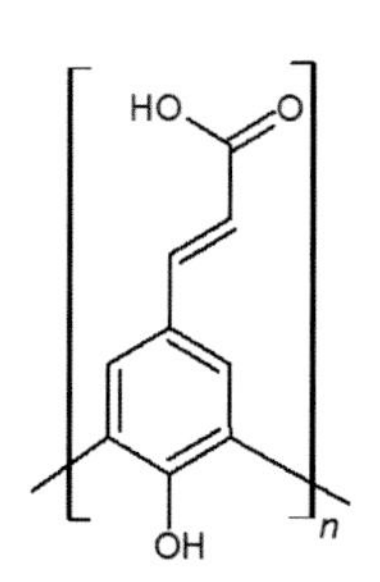

Allomelanin monomer

Black oat

Structures of animal and plant melanins[284]

The descriptive name 'black' is attached to species and genera from across the plant kingdom, signified by the many Linnaean terms related to the Latin *niger –gra –grum*[285]. The genus *Nigella* (love-in-a-mist), from the diminutive of *niger*, is named for its tiny black seeds. The seed coat of *Nigella* owes its colour to melanin[286], the general term for polymers formed from oxidation products of the amino acid tyrosine. At least four types of melanin are recognised. Eumelanin, pheomelanin and neuromelanin are found in animals and determine the colours of skin, hair, feathers, eyes, lips, nerve cells and the inks of squid and octopus. It's surprising how frequently one encounters bald statements such as 'Unlike plants, the animal kingdom possesses a pigment, melanin, that allows for a true black'[287]. In fact melanins of plant origin, referred to as allomelanins, have been known for a long time[288]. As well as *Nigella*, several other species have dark seed coats containing melanin: for example black oat (*Avena strigosa*[289]) and sweet olive (*Osmanthus fragrans*[290]). It's been argued[291] that melanins fulfil the same roles in animals as flavonoids (and, presumably, allomelanins) do in plants, namely as antioxidant protectants against environmental stresses (eg UV light), and as signals ensuring reproductive fitness. Darkening of the infected tissue is a more or less universal plant response to attack by fungi and other disease organisms[292]. Both the host and the pathogen may deploy melanin for attack and defence[293]. The biosynthetic pathway for allomelanins is poorly understood. The amino acid tyrosine is believed to be the main precursor, and DOPA an early intermediate[294], though phenolics including caffeic, chlorogenic, protocatechuic or gallic acid may also feature in the pathway[295]. Studies of black-seeded *Brassica* species suggest oxidation by laccase, tyrosinase or peroxidase enzymes is a common feature of proanthocyanidin, lignin and melanin biosynthesis[296].

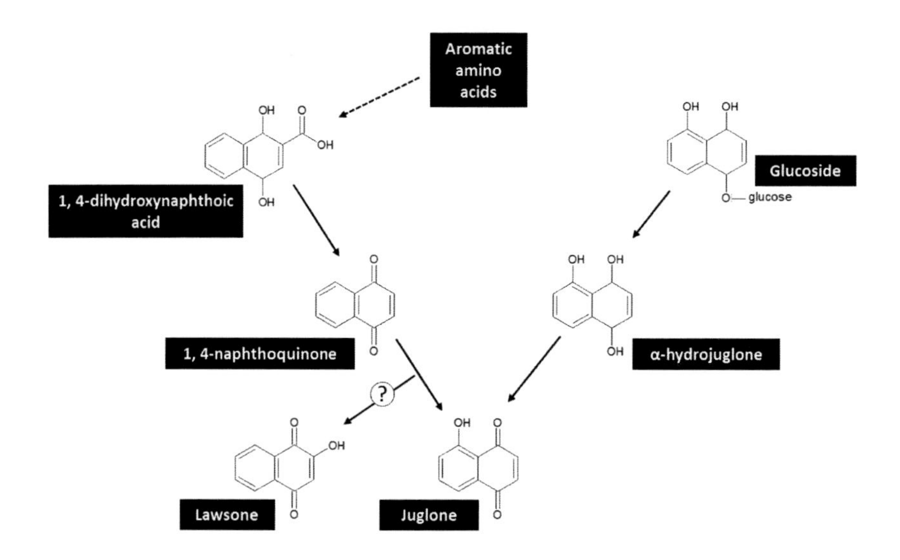

Proposed pathway of juglone and lawsone synthesis[297]

The origin of the pigmentation that gives black walnut (*Juglans nigra*) its name is juglone[298]. Walnut is a famously allelopathic plant - it secretes a melange of chemicals, including juglone and related compounds, that severely inhibit neighbouring competitor species[299]. Likewise DOPA and allomelanins have potent allelochemical properties[300]. Juglone is a naphthoquinone and is a product of the same metabolic network that produces the prenylquinone vitamins K_1 and E and the quinone electron carriers of chloroplasts and mitochondria[301]. Juglone is generally stored as a glycoside, which is readily hydrolysed on disruption of the cell, or by enzymes in the soil, to yield α–hydrojuglone, which in turn is chemically oxidised to the active compound[302]. Handling walnut fruits or wood can stain the skin yellow, brown or black, the result of the generation of free radicals and conjugation of juglone and other quinones with dermal macromolecules. There is a long history of juglone use as a hair dye, wood stain and ink. Its bioactivity has attracted much biomedical interest as a potential treatment for various cancers[303]. Juglone is toxic to fish[304]. Lawsone, the active principle of henna dyes used for hair colouring and tattoos, is structurally related to juglone[305]. Naphthoquinones are among the constituents responsible for the dark colours of heartwood[306] in walnut and the intense blackness of ebony and grenadilla[307], the characteristic woods used in the manufacture of musical instruments[308].

Physiological darkening of plant tissues[309]

When Robert Burton[310] designated 'an old...dizzard, that hath one foot in his grave' as *acherontic*, he gave us a useful word to describe an important, but relatively unregarded, phase in the lifetimes of living organisms and their parts. This is the twilight zone between viability and death, in which biological chemistry and biophysics give way to necrochemistry and entropy. The transition is of particular significance in plants, which are made of disposable parts and deploy acherontic change as an intrinsic physiological process essential for development and adaptation[311]. And it's a characteristic of the acherontic phase that colours morph into the browns, greys and blacks of the post-mortem state[312]. We see this in the appearance of wood, particularly heartwood, which is a tissue made of dead cells; in leaf litter; in diseased tissues where the acherontic reaction has a cauterising function; in the post-harvest deterioration of fruits and vegetables; and in the terminal development of all manner of plant structures such as bark, spines, scales and seed coats[313]. The fact that acherontic change is a predictable and reproducible event in the life of a tissue or organ suggests that it's genetically programmed, but there's a wide-ranging debate to be had about what this really means[314]. For present purposes, it's enough to identify a pigmentation-related common biochemical factor across the spectrum of acherontic processes, namely the activity of enzymes of the polyphenol oxidase group. Polyphenol is a general term for phenylpropanoid-like derivatives with chemical structures that include a number of aromatic and hydroxyl (-OH) groups[315]. Polyphenol oxidases (PPOs) are widely distributed copper-containing plant enzymes that catalyse the reaction between di- or tri-hydroxy phenols and molecular oxygen to produce quinones. Quinone products are typically brown (they account for the appearance of overripe bananas, bruised apples and deteriorating post-harvest vegetables), and such quinones react with other molecules to form melaninoids, tanninoids and suchlike dark insoluble materials. As well as their role in cell death[316], defence and other acherontic processes[317], enzymes of the PPO family contribute to the synthesis of less sombre pigments: tyrosinase (betalain biosynthesis) and aurone synthase (synthesis of the yellow aurone pigments of *Antirrhinum*) are functionally and structurally PPO homologues[318].

Spectral sensitivities of animal vision, and appearance of *Hemerocallis* flower illuminated at different wavelengths[319]

This book is mostly concerned with the colours perceived by the normal trichromatic (RGB) human eye. But there are many other eyes that engage with spectral signatures in the environment. Birds and insects are tetrachromatic, sensitive to reflectance at UV wavelengths[320], and snakes can 'see' into the infrared. Dogs and cats (and human deuteranopes) are red-green colour blind. Vision is more than the passive stimulation of photoreceptors in the eye. It requires a positive act of cognition to make sense of visual stimuli[321]. In the 'language' of Bee World, there will be 'words' for the 'colours' beyond the short wavelength limit of human vision, but we can never know what they are. As Wittgenstein tells us 'If a lion could talk, we could not understand him'[322]. The best we are able to do is to reveal the hidden patterns, such as floral honey or nectar guides and pollinator targets, by observing them under UV illumination, where they commonly appear as dark patches, the consequence of high phenylpropanoid concentrations[323]. In a way, plants have eyes. They are sensitive to the visible light environment, through chlorophyll, carotenoids and phytochrome that drive photosynthesis and photomorphogenesis; but they are equipped with photoreceptors that respond to wavelengths at the blue-violet-UV end of the spectrum too. Cryptochromes, phototropins and zeitlupes perceive and transduce short-wavelength signals from the light environment through protein-bound photoreactive flavin and pterin groups[324]. They are receptors that participate in the control of growth, the biological clock, flowering, stomatal opening and organelle movements, and they interact with the phytochrome network[325]. A specific photoreceptor, UVR8, regulates responses that protect against damage by ionizing UVB radiation (λ <315 nm) through activation of a number of defence genes, including those for pathways that synthesise phenylpropanoid sunblockers[326]. In this case, specific tryptophan residues in the UVR8 protein are the photoreceptive structures. An emerging understanding of plant responses to the long wavelength (IR) end of the spectrum has identified phytochrome as a thermosensor as well as a photosensor[327]. Plants are also attuned to temperature through the changing fluidity of their cell membranes as membrane lipids undergo a physical transition from the gel state in the cold to liquid crystalline when they warm up[328].

Iridescent leaves of *Trichomanes*, and pubescent edelweiss inflorescence[329]

And now we briefly turn to the question of biological colour without pigments. Glover and Whitney[330] distinguish between pigment colour and structural colour and describe study of the latter in plants as the poor relation. Iridescence, which enkindles the spectacular brilliance of bird plumage and butterfly wings, is a form of structural colour that has long been a source of fascination for biologists and physicists alike[331]. It's a natural example of the diffraction grating, widely used as an alternative to the classical prism beam-splitter in optical instrumentation[332], and familiar from the rainbow sheen on the surfaces of compact discs and soap bubbles[333]. David Lee's classic work on plant colour, *Nature's Palette*[334], gives an absorbing personal account of research into plant iridescence. His fascination with the blue foliage of understorey rainforest plants (mostly ferns and their allies) led him to identify multilayered light-scattering structures in many of these species. As for the function of foliar iridescence, it awaits a comprehensive explanation. Suggestions include enhanced photosynthetic light capture, defence against photodamage, and herbivore deterrence[335]. The floral epidermal cells of several species form light-scattering surface structures and some have the characteristics of diffraction gratings[336]. It has been proposed that signals from such periodic and reflective features are attractive to pollinators, but this view is disputed, and it may be that there's a compromise between the confusing effect of iridescence on insect vision and the beneficial influence of iridescence on object detectability[337]. Other surface structures that change the optical properties of plant parts include hairs and waxes. Epicuticular wax imparts a white bloom, with a variety of functions across a wide range of plant species, discussed in a recent comprehensive review[338]: Many plants of alpine, arid, coastal and dune environments are covered with white trichomes (epidermal hairs). Hairiness is an adaptation for controlling leaf heat balance and photon interception, protecting against photoinhibition and UVB damage to leaf photochemistry, modulating leaf water relations and wettability, and mimicry to deter herbivore damage[339]. The surface hairs of edelweiss (*Leontopodium*) are of particular interest to physicists because they absorb the harmful UV levels in the high altitude alpine habitat by virtue of their near-perfect photonic crystal fibre optic structure[340].

Elenydd[341]

Photosynthetic pigments stand at the gateway through which carbon enters the great biogeochemical cycle of organic turnover. Pigments are there, too, at the end of organic carbon's journey through the biosphere: the interaction between light and the pigments of end-of-life, discarded (post-acherontic, you might say) plant material makes an essential contribution to carbon storage, transformation and loss in terrestrial ecosystems[342]. A striking reciprocal relationship is often observed between biotic litter decomposition and photodegradation, under the influence of water availability, fire, human-mediated land use and other environmental factors[343]. The degree to which plant litter absorbs UV and visible light is determined by its chemical composition, with phenolics playing a leading role. The UV-absorbing properties of lignin are of particular significance, since lignin accounts for around 30% of terrestrial post-acherontic carbon sequestered annually and is a bottleneck in the geochemical turnover cycle[344]. Light-mediated bleaching of lignified tissue makes cellulose available, resulting in saccharification (sweetening) of plant litter and the promotion of microbial population growth[345]. Thus pigment action and quietus in the terminal phase of the lifecycles of plants and their parts have ecosystem-wide influence.

Which leads me to conclude with an example from my own environment. Until well into the 20th century, travellers to the central uplands of Wales found a region of sparse population, no roads except sheep-walks and drovers' tracks, and mile upon mile of treeless moorland[346]. They called it the Welsh Desert. The ancient Welsh name for this region is Elenydd, and its part in archaeology, myth, legend and literature extends back to the earliest period in European history[347]. The vegetation is dominated by *Molinia caerulea,* a coarse tussock species that is unusual among grasses in being deciduous[348]. Every year at the end of the growing season, *Molinia* foliage senesces, eliminates its photosynthetic pigments, exports its nutrients to the extensive rhizomatous root system, abscinds and forms vast mats of photobleached leaves[349]. Colour drains away, leaving a landscape that, under the soft light of late season, resembles nothing so much as a vista of sand dunes receding towards a distant horizon. A true Welsh desert indeed.

And with that, the Tale of the Three Little Pigments comes to an end.

NOTES AND SOURCES

[277] Remains of the seed-head of *Lotus*. Public domain image under Creative Commons CC0, from https://pixabay.com/en/lotus-canopy-lotus-black-dry-hole-1752445/ [accessed 31 May 2018].

[278] https://youtu.be/yLF82_esjec [accessed 27 June 2018].

[279] Nemcsics A, Caivano JL. 2015. Color order systems. *Encyclopedia of Color Science and Technology* (ed MR Luo) pp 1-16. NY: Springer.

[280] Gage J. 1999. *Colour and Meaning: Art, Science and Symbolism*. London: Thames and Hudson

[281] Nemcsics and Caivano (2015)

[282] According to the influential theory proposed by Berlin and Kay (1969), different languages divide up colour space with respect to a set of basic colour terms, which are acquired in a specific sequence. Thus a 'Stage I' language will have terms for black and white. At 'Stage II' a word for red is adopted. Yellow and green appear at 'Stage III' and 'Stage IV'. Blue is introduced in 'Stage V', and brown in 'Stage VI'. It's not until 'Stage VII' that words for pink, purple, orange, and grey are acquired. The soundness of this scheme continues to be vigorously debated, but for our purposes it's striking the way it maps onto the physiological landscape of plant pigmentation (Berlin B, Kay P. 1969. *Basic Color Terms*. Berkeley: University of California Press).

[283] Katsushika Hokusai, 19th century artist and printmaker, creator of the famous print 'The Great Wave off Kanagawa', wrote in a treatise on colour 'There is a black which is old and a black which is fresh. Lustrous black and matt black, black in sunlight and black in shadow'. And there's this, by Paolo Giordano: 'Mattia was right: the days had slipped over her skin like a solvent, one after the other, each removing a very thin layer of pigment from her tattoo, and from both of their memories. The outlines, like the circumstances, were still there, black and well delineated, but the colors had merged together until they faded into a dull, uniform tonality, a neutral absence of meaning' (Giordano P. 2009. *The Solitude of Prime Numbers*. London: Black Swan).

[284] Varga M, Berkesi O, Darula Z, May NV, Palágyi A. 2016. Structural characterization of allomelanin from black oat. Phytochemistry 130: 313-320. GFDL image of *Avena sativa* by Rasbak (https://commons.wikimedia.org/wiki/File:Avena_sativa_black_oat,_zwarte_haver (1).jpg#file [accessed 25 May 2018]). For oat allomelanin, n = 3-9.

[285] Gledhill D. 2008. *The Names of Plants*. Cambridge: University Press.

[286] Al-Mufarrej SI, Hassib AM, Hussein MF. 2006. Effect of melanin extract from black cumin seeds (*Nigella sativa* L.) on humoral antibody response to sheep red blood to cells in albino rats. Journal of Applied Animal Research 29: 37-41.

[287] St Clair K. 2016. *The Secret Lives of Colour*. London: Murray.

[288] Andrews RS, Pridham JB. 1967. Melanins from DOPA-containing plants. Phytochemistry 6: 13-18. The browning that occurs in fruits and vegetables post-harvest has been attributed to the formation of 'phytomelanins': Mesquita VLV, Queiroz C. 2013. Enzymatic browning. In: Eskin NAM, Shahidi F (eds) *Biochemistry of Foods*. London: Academic Press, London, pp

387–418. The brown colour of chocolate that develops after curing, drying and roasting cocoa beans is due to 'melanoidins': Tannenbaum G. 2004. Chocolate: a marvelous natural product of chemistry. Journal of Chemical Education 81: 1131-1135.
[289] Varga et al. (2016).
[290] Wang H, Pan Y, Tang X, Huang Z. 2006. Isolation and characterization of melanin from *Osmanthus fragrans* seeds. LWT-Food Science and Technology 39: 496-502.
[291] Carletti G, Nervo G, Cattivelli L. 2014. Flavonoids and melanins: a common strategy across two kingdoms. International Journal of Biological Sciences 10: 1159.
[292] Rubin BA, Artsikhovskaya EV. 1964. Biochemistry of pathological darkening of plant tissues. Annual Review of Phytopathology 2: 157-178.
[293] Howard RJ, Ferrari MA. 1989. Role of melanin in appressorium function. Experimental Mycology 13: 403-418.
[294] Andrews and Pridham (1967); Soares AR, Marchiosi R, Siqueira-Soares RD, Barbosa de Lima R, Dantas dos Santos W, Ferrarese-Filho O. 2014. The role of L-DOPA in plants. Plant Signaling and Behavior 9: e28275. Allomelanins, alkaloids and betalain pigments are likely to have a common metabolic origin. The KEGG map covering reactions from tyrosine to eu- and pheo- (but not yet allo-) melanins is at www.genome.jp/kegg-bin/ show_pathway?map00350+C17937 [accessed 23 May 2018].
[295] Varga et al. (2016).
[296] Yu CY. 2013. Molecular mechanism of manipulating seed coat coloration in oilseed *Brassica* species. Journal of Applied Genetics 54: 135-145.
[297] Strugstad M, Despotovski S. 2013. A summary of extraction, synthesis, properties, and potential uses of juglone: a literature review. Journal of Ecosystems and Management 13: 1-16; McCoy RM, Ye YJ, Widhalm JR. 2017. Elucidation of juglone synthesis in black walnut. FASEB Journal 31(1 Supplement): lb212.
[298] Strugstad and Despotovski (2013).
[299] Soderquist CJ. 1973. Juglone and allelopathy. Journal of Chemical Education 50: 782–783.
[300] Soares et al. (2014).
[301] Widhalm JR, Rhodes D. 2016. Biosynthesis and molecular actions of specialized 1, 4-naphthoquinone natural products produced by horticultural plants. Horticulture Research 3: 16046.
[302] Strugstad and Despotovski (2013).
[303] Thakur A. 2011. Juglone: A therapeutic phytochemical from *Juglans regia* L. Journal of Medicinal Plants Research 5: 5324-5330.
[304] Marking LL. 1970. Juglone (5-hydroxy-1, 4-naphthoquinone) as a fish toxicant. Transactions of the American Fisheries Society 99: 510-514. There are reports of poachers killing fish by casting walnut husks into the water.
[305] The practice of tattooing dates back millennia (Deter-Wolf et al. 2016). The upsurge of interest in body art in recent decades (Friedman 2015) has created a demand for safe pigments, avoiding the use of harmful chemical additives, notably para-phenylenediamine (PPD; Önder 2003). One such temporary tattoo dye (not without its undesirable side-effects) is based on the sap from the immature fruit of jagua (*Genipa americana*), which contains geniposide and its bioactive derivative genipin. Genipin itself is colourless, but reacts spontaneously with proteins and amino acids to form blue–black pigments (Bircher et al. 2017); Deter-Wolf A, Robitaille B, Krutak L, Galliot S. 2016. The world's oldest tattoos. Journal of Archaeological Science 5: 19–24; Friedman AF. 2015. *The World Atlas of Tattoo.*

London: Thames and Hudson; Önder M. 2003. Temporary holiday tattoos may cause lifelong allergic contact dermatitis when henna is mixed with PPD. Journal of Cosmetic Dermatology 2: 126–130; Bircher AJ, Sigg R, Scherer Hofmeier K, Schlegel U, Hauri U. 2017. Allergic contact dermatitis caused by a new temporary blue–black tattoo dye – sensitization to genipin from jagua (*Genipa americana* L.) fruit extract. Contact Dermatitis 77: 374-378.

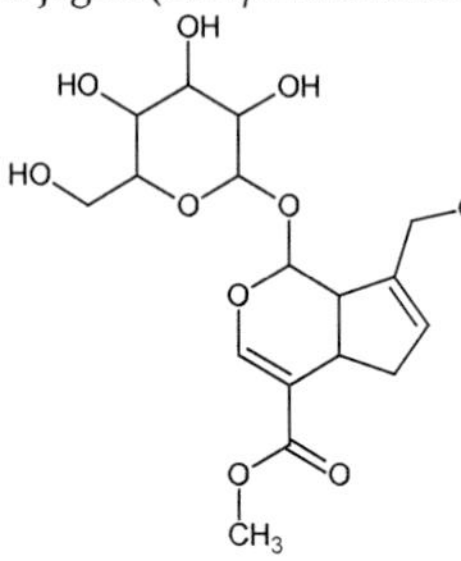

Geniposide

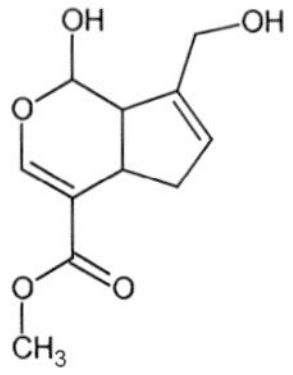

Genipin

[306] Yazaki Y. 2015. Wood colors and their coloring matters: a review. Natural Product Communications 10: 505-512; Celedon JM, Bohlmann J. 2017. An extended model of heartwood secondary metabolism informed by functional genomics. Tree Physiology 38: 311-319.

[307] Brown AG, Lovie JC, Thomson RH. 1965. Ebenaceae extractives. 1. Naphthalene derivatives from Macassar ebony (*Diospyros celebica* Bakh). Journal of the Chemical Society 2355–2361; Yoshihira K, Tezuka M, Takahashi C, Natori S. 1971. Four new naphthoquinone derivatives from *Diospyros* spp. Chemical and Pharmaceutical Bulletin 19: 851–854.

[308] The tailpiece and fingerboard of a good violin will be made of ebony. Over-exploitation of ebony in tropical countries has designated it a protected wood species under the Convention on International Trade in Endangered Species of Wild Fauna and Flora (CITES). There are cases where musicians have had their violins confiscated at customs when entering a country signed up to CITES. Ebony may be imported only if its legal provenance can be demonstrated, and musicians, instrument manufacturers and dealers have, often unwittingly, found themselves in breach of international law. Grenadilla wood (used for clarinets, oboes and flutes) became CITES listed in 2017.

[309] The musical instrument is a zurna, a wind instrument of middle Eurasian origin, made from black grenadilla heartwood (public domain CC0 image, https://commons.wikimedia.org/wiki/File:Zurna,_Boru.jpg). Other pictures are author's own work.

[310] Burton R. 1621. *The Anatomy of Melancholy, What it is: With all the Kinds, Causes, Symptomes, Prognostickes, and Several Cures of it. In Three Maine Partitions with their several Sections, Members, and Subsections. Philosophically, Medicinally, Historically, Opened and Cut Up.* 1638 edition. Oxford: Henry Cripps.

[311] Thomas H. 2016. *Senescence*. Aberystwyth: Thomas.

[312] The post-acherontic fate of a whole flora, moving through colour space in the direction of (literally) pitch black, has named a geological era: the Carboniferous. See Locatelli ER. 2014. The exceptional preservation of plant fossils: a review of taphonomic pathways and biases in the fossil record. The Paleontological Society Papers 20: 237-258.

[313] And galls (I'm writing this on Oak-Apple Day), which have a history of diverse uses, including medicine, ink manufacture, tanning, dyeing and even food and fuel (Fagan MM. 1918. The uses of insect galls. The American Naturalist 52: 155-176).

[314] Tangled up in this debate is the question of programmed cell death, a term that is bandied about, rather carelessly in my view. This is what we (Thomas et al. 2003) wrote about this in a discussion paper: 'The idea of genetic programming has to be applied with caution to events that occur in cells in the intermediate state between living and dead…Of course, post-mortem changes will certainly have a purposeful, non-random appearance. The structures of (acherontic) cells, membranes, nucleic acids or proteins will collapse in a semi-reproducible fashion because they are built the way they are' (Thomas H, Ougham HJ, Wagstaff C, Stead AJ. 2003. Defining senescence and death. Journal of Experimental Botany 54: 1127-1132).

[315] Quideau S. 2009. Why bother with polyphenols? www.groupepolyphenols.com/the-society/why-bother-with-polyphenols/ [accessed 21 May 2018].

[316] Araji S, Grammer TA, Gertzen R, Anderson SD, Mikulic-Petkovsek M, Veberic R, Phu ML, Solar A, Leslie CA, Dandekar AM, Escobar MA. 2014. Novel roles for the polyphenol oxidase enzyme in secondary metabolism and the regulation of cell death in walnut. Plant Physiology 164: 1191-1203.

[317] Jukanti A. 2017. *Polyphenol Oxidases (PPOs) in Plants*. Singapore: Springer

[318] Sullivan ML. 2015. Beyond brown: polyphenol oxidases as enzymes of plant specialized metabolism. Frontiers in Plant Science 5: 783.

[319] Images of daylily flower under visible, ultraviolet and infrared light by David Kennard (under CC-BY-SA 3.0; www.davidkennardphotography.com/ [accessed 30 May 2018]).

[320] Primack RB. 1982. Ultraviolet patterns in flowers, or flowers as viewed by insects. Arnoldia 42: 139–146; Papiorek S, Junker RR, Alves-dos-Santos I, Melo GA, Amaral-Neto LP, Sazima M, Wolowski M, Freitas L, Lunau K. 2016. Bees, birds and yellow flowers: pollinator-dependent convergent evolution of UV patterns. Plant Biology 18: 46-55. The FReD database is a compilation of the absorbance and reflectance spectra of over 2000 flowers in relation to insect vision (www.reflectance.co.uk//index.php [accessed 30 May 2018]).

[321] Zajonc A. 1993. *Catching the Light: The Entwined History of Light and Mind*. Oxford: University Press.

[322] Wittgenstein L. 1953. *Philosophical Investigations*. New York: Macmillan.

[323] For example, the distribution of sinapoyl glucoside determines patterning in the petals of *Brassica* spp.

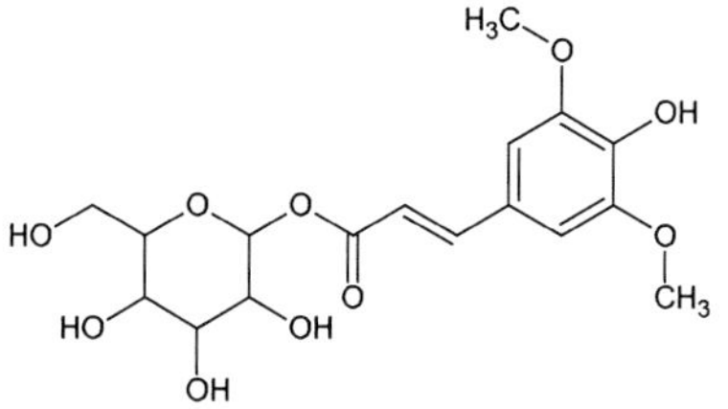

Sinapoyl-D-glucose

Brock MT, Lucas LK, Anderson NA, Rubin MJ, Cody Markelz RJ, Covington MF, Devisetty UK, Chapple C, Maloof JN, Weinig C. 2016. Genetic architecture, biochemical underpinnings and ecological impact of floral UV patterning. Molecular Ecology 25: 1122-1140.

[324] Flavins are yellow pigments with a range of physiological functions. Signal perception and transduction by blue/UV-sensitive receptors is initiated by photoreduction of the flavin group bound to the receptor protein. Pterins are members of a family of bioactives that contribute to pigmentation in insects, reptiles, fish and birds.

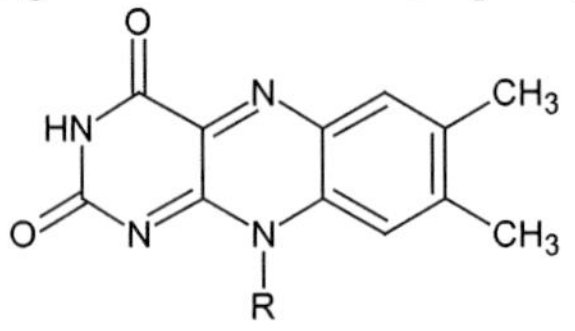

Flavin 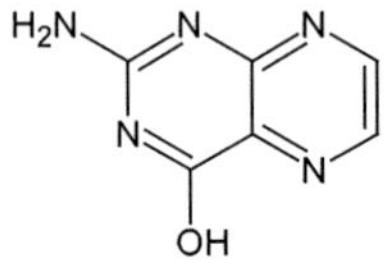　　　Pterin

[325] de Wit M, Galvão VC, Fankhauser C. 2016. Light-mediated hormonal regulation of plant growth and development. Annual Review of Plant Biology 67: 513-537.

[326] Jenkins GI. 2014. The UV-B photoreceptor UVR8: from structure to physiology. Plant Cell 26: 21-37.

[327] Delker C, van Zanten M, Quint M. 2017. Thermosensing enlightened. Trends in Plant Science 22: 185-187.

[328] Bahuguna RN, Jagadish KS. 2015. Temperature regulation of plant phenological development. Environmental and Experimental Botany 111: 83-90.

[329] The iridescent blue foliage of the rain forest filmy fern *Trichomanes elegans* (Image reproduced under CC BY-ND 2.0, from www.flickr.com/photos/lungfish2000/38062253582/in/photostream/ [accessed 3 June 2018]). Hairy bracts of edelweiss inflorescence (Image by Michael Schmid under CC BY-SA 2.0, https://commons.wikimedia.org/wiki//l/File:Leontopodium_alpinum_detail.jpg [accessed 3 June 2018]).

[330] Glover BJ, Whitney HM. 2010. Structural colour and iridescence in plants: the poorly studied relations of pigment colour. Annals of Botany 105: 505–511.

[331] Doucet SM, Meadows MG. 2009. Iridescence: a functional perspective. Journal of The Royal Society Interface 6(Suppl 2): S115-132.

[332] Iridescent materials are increasingly used for furniture, home accessories, clothing architecture and all kinds of designed objects (www.fastcodesign.com/3068789/the-weird-and-fascinating-story-behind-designs-iridescence-craze [accessed 3 June 2018}). In this connexion, it's worth mentioning a remarkable example of synthetic structural colour (or, more accurately, lack thereof), Vantablack, a proprietary material made from carbon nanotubes, from which more than 99.9% of the light that enters never re-emerges (Jackson JJ, Puretzky AA, More KL, Rouleau CM, Eres G, Geohegan DB. 2010. Pulsed growth of vertically aligned nanotube arrays with variable density. ACS Nano 4: 7573-7581). Called the blackest known artificial substance, it is finding a variety of practical applications: for example, it was used by the architect Asif Khan for the Hyundai pavilion at the 2018 South Korean Winter Olympics, to create 'the darkest building on Earth' (http://tinyurl.com/yauc5ta3 [accessed 3 June 2018]). Not for the first time, this technological development was anticipated years previously by Douglas Adams: 'It was a ship of classic, simple design, like a flattened salmon, twenty yards long, very clean, very sleek. There was just one remarkable thing about it. "It's so ... black!" said Ford Prefect, "you

can hardly make out its shape ... light just seems to fall into it!'" (Adams D. 1980. *The Restaurant at the End of the Universe*. London: Pan).
[333] The best explanation of how a diffraction grating works that I know of can be found in Feynmann R. 1990. *QED - the Strange Theory of Light and Matter*. London: Penguin.
[334] *Nature's Palette* tells engagingly of the author's encounters with pigments, particularly those of the vivid and lush floras of the subtropical and tropical areas he has worked and lived in. As an inhabitant of the sombre and chilly temperate climes of Northern Europe, I can't match David's kaleidoscopic experiences. Someone once said that people of the sun-blessed south dance in the open air and drink wine while those of the grey and rainy north hide in cave-like houses and drink beer. I can't argue with that, and can only hope this contrast gives my book a perspective on colour in the environment that complements that of *Nature's Palette*. (Lee D. 2007. *Nature's Palette. The Science of Plant Color*. Chicago University Press).
[335] Thomas KR, Kolle M, Whitney HM, Glover BJ, Steiner U. 2010. Function of blue iridescence in tropical understorey plants. JRS Interface 7: 1699–1707.
[336] Glover and Whitney (2010).
[337] Kooi CJ, Wilts BD, Leertouwer HL, Staal M, Elzenga JT, Stavenga DG. 2014. Iridescent flowers? Contribution of surface structures to optical signaling. New Phytologist 203: 667-673; Whitney HM, Reed A, Rands SA, Chittka L, Glover BJ. 2016. Flower iridescence increases object detection in the insect visual system without compromising object identity. Current Biology 26: 802-808.
[338] Some important roles for waxes: mechanical stability; spectral reflection and absorption; enhanced water relations; adhesion e.g. insect trapping; aero- and hydro-dynamic efficiency (Barthlott W, Mail M, Bhushan B, Koch K. 2017. Plant surfaces: structures and functions for biomimetic innovations. Nano-Micro Letters 9(2):23).
[339] Lev-Yadun S. 2016. *Defensive (anti-herbivory) Coloration in Land Plants*. Cham: Springer; Bickford CP. 2016. Ecophysiology of leaf trichomes. Functional Plant Biology 43:807-814.
[340] Vigneron JP, Rassart M, Vértesy Z, Kertész K, Sarrazin M, Biró LP, Ertz D, Lousse V. 2005. Optical structure and function of the white filamentary hair covering the edelweiss bracts. Physical Review E 71(1): 011906.
[341] Photograph reproduced from http://tinyurl.com/y8jxswdf (under Creative Commons CC BY-SA [accessed 12 June 2018]).
[342] Cornwell WK, Cornelissen JH, Amatangelo K, Dorrepaal E, Eviner VT, Godoy O, Hobbie SE, Hoorens B, Kurokawa H, Pérez-Harguindeguy N, Quested HM. 2008. Plant species traits are the predominant control on litter decomposition rates within biomes worldwide. Ecology Letters 11: 1065-1071.
[343] Austin AT. 2011. Has water limited our imagination for aridland biogeochemistry?. Trends in Ecology and Evolution 26: 229-235.
[344] Austin AT, Ballaré CL. 2010. Dual role of lignin in plant litter decomposition in terrestrial ecosystems. Proceedings of the National Academy of Sciences 107: 4618-4622.
[345] Austin AT, Méndez MS, Ballaré CL. 2016. Photodegradation alleviates the lignin bottleneck for carbon turnover in terrestrial ecosystems. Proceedings of the National Academy of Sciences 113: 4392-4397.
[346] It remains to this day a remote place that's difficult to get into and out of.
[347] Elenydd is the theatre in which the classical works of Welsh literature have been played out: Graves R. 1948. *The White Goddess*. London: Faber and Faber (2010 edition); Francis M.

2018. *The Mabinogi*. London: Faber and Faber. For the historical significance of Elenydd, see www.walesonline.co.uk/news/wales-news/welsh-history-month-pumlumon-elenydd-2046978 [accessed 25 June 2018].
[348] Taylor K, Rowland AP, Jones HE. 2001. *Molinia caerulea* (L.) Moench. Journal of Ecology 89: 126-144.
[349] When foliar pigments, nutrients and water have departed, what is left is lignocellulose - paper, in other words. I imagine Elenydd to be a paper landscape on which is written, in faded, almost invisible, letters a deep history of plants, pigments and people. The great poet of Elenydd in the modern era was RS Thomas. Here is 'The Moor' (from *Pietá*, 1966):

> It was like a church to me.
> I entered it on soft foot,
> Breath held like a cap in the hand.
> It was quiet.
> What God there was made himself felt,
> Not listened to, in clean colours
> That brought a moistening of the eye,
> In a movement of the wind over grass.
>
> There were no prayers said. But stillness
> Of the heart's passions -- that was praise
> Enough; and the mind's cession
> Of its kingdom. I walked on,
> Simple and poor, while the air crumbled
> And broke on me generously as bread.

INDEX

ABOUT

'Que sais-je?' wrote Michel de Montaigne in *Essais* (1580). 'What do I know?' Well, I suppose after a career in research on senescence and grasses, I ought to know something about those subjects; and so, in the digressive, anecdotal (and possibly self-indulgent) spirit of Montaigne, I wrote *Senescence* in 2016 and *The War Between Trees and Grasses* in 2017. I expected the tank would be empty once they'd been published. But another subject I'm familiar with started to tug at my sleeve and the next thing I knew, I had the final part of a trilogy on my hands, an account of colour as the essence of plants and their place in our world.

Although across the span of human history we have admired, exploited, dismantled, rebuilt, modified and interrogated them, plant pigments aren't there for our benefit. Colour may be intrinsic to a coevolutionary contract between a plant and another organism, but a pigment may equally have a function in a plant's fitness regime for which colour is merely incidental or irrelevant. Even bearing these truths in mind, it's not easy - maybe impossible - to think like a plant as well as a person, but that's what I've tried to do. It's what makes botany such a richly transcendent discipline. In this sense, there's a grain of truth in the words of Rick Martinez: 'But, I mean, it's only botany. It's not real science' (*The Martian*, directed by Ridley Scott, 2015).

Most of the people who, largely unknowingly, helped to make this book possible. are acknowledged in NOTES AND SOURCES. The framework of this book took shape when Helen Ougham and I wrote 'Y ddeilen hon: natur, gwreiddiau a phwrpasau lliwiau dail', for the Welsh language science journal Gwerddon. As ever, I'm grateful for Helen's editorial interventions, and her patience during all the hours when I disappeared to the cwtch in the company of the Three Little Pigments. Sarah Lennon kindly dealt with permissions on my behalf - thanks, Sarah. And I would not have been able to write and publish this book and its predecessors without the continued generous support of Aberystwyth University and the New Phytologist Trust.

In the Tale of the Three Little Pigments, no-one lives happily ever after: the Big Bad Wolf of Destiny finally blows the house down. In this reflective mood, and having begun this book with the words of Chaucer's Wife of Bath, it seems fitting to end with Chaucer's pensive farewell to his *Troilus and Criseyde.*

Go, litel book, go litel myn tragedie

Howard Thomas,
Aberystwyth and Wye
August 2018

Howard Thomas was born and educated in Wales and, after a career in scientific research including visiting professorships at Universities in Japan, the United States and Switzerland, he is now emeritus Professor of Biology at Aberystwyth University. He has published extensively on the genetics and physiology of plant development and has a special interest in the science-humanities connection. He is a Fellow of the Learned Society of Wales, a Trustee of the New Phytologist and a co-author of *The Molecular Life of Plants* (2012, Wiley) and *Food and the Literary Imagination* (2014, Palgrave). His most recent books are *Senescence* (2016) and *The War Between Trees and Grasses* (2017). He is also a devout jazz musician, and is author of *20 Steps to Jazz Keyboard Harmony* (2015, Smashwords).

www.sidthomas.net/wp